CONTENTS

INTERPOLATION

CURVE FITTING

NUMERICAL DIFFERENTIATION

NUMERICAL INTEGRATION

PROLOGUE

1 HISTORICAL

Although some may regard Numerical Analysis as a subject of recent origin, this in fact is not so. In the first place, it is concerned with the provision of results in the form of numbers, which no doubt were in use by very early man. More recently, the Babylonian and ancient Egyptian cultures were noteworthy for numerical expertise, particularly in association with astronomy and civil engineering. There is a Babylonian tablet dated approximately 2000 B.C. giving the squares of the integers 1-60, and another which records the eclipses back to about 750 B.C. The Egyptians dealt with fractions, and even invented the *method of false position* for the solution of algebraic equations (see Step 8).

It is probably unnecessary to point out that the Greeks produced a number of outstanding mathematicians, many of whom provided important numerical results. In about 220 B.C. Archimedes gave the result

$$3\tfrac{10}{71} < \pi < 3\tfrac{1}{7}.$$

The iterative procedure for $\sqrt{a}$ involving $\tfrac{1}{2}(x_n + \dfrac{a}{x_n})$ and usually attributed to Newton (see Step 10) was in fact used by Heron the elder in about 100 B.C. The Pythagoreans considered the numerical summation of series, and Diophantus in about 250 A.D. gave a process for the solution of quadratic equations.

Subsequently, progress in numerical work occurred in the Middle East. Apart from the development of the modern arithmetical notation commonly referred to as Arabic, tables of the trigonometric functions sine and tangent were constructed by the tenth century. Further east, in India and China, there was parallel (although not altogether separate) mathematical evolution.

In the West, the Renaissance and scientific revolution involved a rapid expansion of mathematical knowledge, including the field of Numerical Analysis. Such great names of mathematics as Newton,

Euler, Lagrange, Gauss and Bessel are associated with modern methods of Numerical Analysis, and testify to the widespread interest in the subject.

In the seventeenth century, Napier produced a table of logarithms, Oughtred invented the slide rule, and Pascal and Leibniz pioneered the development of calculating machines (although these were not produced in quantity until the nineteenth century). The provision of such machines brought a revolution in numerical work, a revolution accentuated since the late 1940's by the development of the electronic computer.

The extent of this revolution becomes clear when we consider the improvements in calculating speeds involved; whereas mechanical calculators are about 10 times faster than pencil and paper, modern electronic computers are about 10 million times faster again! New procedures have been and are being developed; computations and data analyses which could not have been contemplated even as a life's work a few decades ago are now solved in hours. The machinery at our disposal is the dominant new feature in the field of Numerical Analysis.

2 NUMERICAL ANALYSIS TODAY

Theoretical science involves the construction of *models* to interpret experimental results, and to predict results for future experimental check. Since these results are often numerical, the applied mathematician attempts to construct a *mathematical model* of a complex situation arising in some field such as physics or economics by describing the important features in mathematical terms. The art of good applied mathematics is to retain only those features essential for useful deductions, for otherwise there is usually unnecessary extra work.

The abstract nature of such a mathematical model can be a real advantage, for it may well be similar to others, previously studied in quite different contexts but whose solutions are known. Occasionally, there may be a formal analytical solution procedure available, but even then it may yield expressions so unwieldy that any subsequent necessary interpretation of the mathematical results is difficult. In many cases, a numerical procedure leading to meaningful numerical results is available *and* preferable. Numerical Analysis remains a branch of mathematics in which such numerical procedures are studied, with emphasis today on techniques for use on automatic digital computers.

There are various main problem areas in Numerical Analysis, including finding roots of non-linear equations, solving systems of linear algebraic equations, correct use of tables, evaluating integrals, solving differential equations, and optimization. Equations involving transcendental functions (e.g. logarithm or sine) often arise in science or engineering, for example, and are usually solved numerically. *Algebraic equations* are common in both science and social science (e.g. the rotation of a set of coordinate axes or the movement of goods in an economy). The solution of *differential equations* is a major requirement in various fields, such as mathematical physics or environmental studies. Since many of these differential equations are non-linear and therefore probably not amenable to analytic solution, their numerical solution is important.

In an introductory text, of course, it is not possible to deal in depth with other than a few basic topics. Nevertheless, we hope by these few remarks to encourage the student not only to view his progress through this book as worthwhile, but also to venture beyond with enthusiasm and success.

3 THIS BOOK

Each main topic treated in the book has been divided into a number of Steps. The first five are devoted to the question of errors arising in numerical work. We believe that a thorough understanding of errors is necessary for a proper appreciation of the art of using numerical methods. The succeeding Steps deal with concepts and methods used in the problem areas of non-linear equations, systems of linear equations, interpolation, differentiation, and integration.

Most of the unasterisked Steps in the book will be included in any first course. The asterisked Steps ('side-steps') include material which the authors consider to be extra, but not necessarily extensive, to a first course. The material in each Step is intended to be an increment of convenient size, perhaps dependent on the understanding of earlier (but not later) un-asterisked Steps. Ideally, the consideration of each Step should involve at least the Exercises, carried out under the supervision of the teacher where necessary. We emphasise that Numerical Analysis demands considerable practical experience, and further exercises could also be valuable.

Within each Step, the concepts and method to be learned are presented first, followed by illustrative examples. The student is then invited to test his immediate understanding of the text by answering two or three Checkpoint questions. These concentrate on salient points made in the Step, induce the student to think about and re-read the text just covered, and may also be useful for revision purposes.

Brief answers are provided at the end of the book for the Exercises set in each Step.

After much consideration, the authors decided not to include computer programs for the various algorithms introduced in the Steps. However, they have provided a set of basic flow-charts in an Appendix. Students will gain much if they study the flow-chart of a method at the same time as they learn it in a Step. If they are familiar with a programming language they should be encouraged to convert at least some of the flow-charts into computer programs, and apply them to the set Exercises.

ERRORS 1
Sources of error

The main sources of error in obtaining numerical solutions to mathematical problems are:

a) the model – its construction usually involves simplifications and omissions;

b) the data – there may be errors in measuring or estimating values;

c) the numerical method – generally based on some sort of approximation;

d) the representation of numbers – e.g. π cannot be represented exactly by a finite number of digits;

e) the arithmetic – frequently errors are introduced in carrying out operations such as addition ($+$) and multiplication ($\times$).

We can pass responsibility for (a) onto the applied mathematician, but the others are not so easy to dismiss. Thus, if the errors in the data are known to lie within certain bounds, we should be able to estimate the consequential errors in the results. Similarly, given the characteristics of the computer, we should be able to account for the effects of (d) and (e). As for (c), when a numerical method is devised it is customary to investigate its error properties.

EXAMPLE
To illustrate the ways in which the above errors arise, let us take the

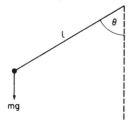

FIGURE 1. Simple pendulum

example of the simple pendulum (see Fig. 1). If air resistance and friction at the pivot are neglected we obtain the simple (non-linear) differential equation

$$ml \frac{d^2\theta}{dt^2} = -mg \sin\theta$$

In introductory mechanics courses† the customary next step is to use the approximation $\sin \theta \approx \theta$ to produce the even simpler (linear) differential equation

$$\frac{d^2\theta}{dt^2} = -\omega^2\theta, \text{ where } \omega^2 = g/l.$$

This has the analytical solution

$$\theta(t) = A \sin \omega t + B \cos \omega t,$$

where A and B are suitable constants.

We can then deduce that the *period* of the simple pendulum (i.e. the smallest positive value of T such that $\theta(t + T) = \theta(t)$) is

$$2\pi/\omega = 2\pi\sqrt{l/g}$$

Up to this point we have encountered only errors of type (a); the other errors are introduced when we try to obtain a numerical value for T in a particular case. Thus both l and g will be subject to *measurement* errors; π must be *represented* as a finite decimal number, the square root must be *computed* (using tables or an iterative process) after *dividing* l by g (which may involve a rounding error), and finally the square root must be *multiplied* by 2π.

Checkpoint

1. What sources of error are of concern to the numerical analyst?
2. Which types of error depend upon the computer used?

---

† In practice one could reduce the type (a) error by using a *numerical method* (see Step 31) to solve the more realistic (non-linear) differential equation
$$\frac{d^2\theta}{dt^2} = -\omega^2\sin\theta$$

EXERCISES

When carrying out the following calculations, notice all the points at which errors of one kind or another arise.

1. Calculate the period of a simple pendulum of length 75 centimetres, given that g is 981 cm/sec^2.

2. The rate of flow of a liquid through a circular hole of diameter d is given by the formula

$$R = C\frac{\pi d^2}{4} \sqrt{2gH},$$

where C is a coefficient of discharge and H is the head of liquid causing the flow. Calculate R for a head of 650 cm, given that $d = 15$ cm and the coefficient of discharge is estimated to be 0·028.

ERRORS 2
Approximation to numbers

Although it may not seem so to the beginner, it is important to examine ways in which numbers are represented.

1 NUMBER REPRESENTATION

We humans normally represent a number in *decimal* (base 10) form, although modern computing machines use binary (base 2) and also hexadecimal (base 16) forms. The arithmetical operation of division often gives a number which does not terminate; the decimal (base 10) representation of $\frac{2}{3}$ is one example. There are also the irrational numbers such as the value of π, which do not terminate. In order to carry out a numerical calculation involving such numbers, we are forced to approximate to them by a representation involving a finite number of *significant figures* (S). For practical reasons (e.g. the size of the back of an envelope or the 'storage' available in a machine), this number is usually quite small.

To five significant figures ($5S$), $\frac{2}{3}$ is represented by 0·66667, π by 3·1416, and $\sqrt{2}$ by 1·4142. None of these is an exact representation, but all are correct to within half a unit of the fifth significant figure. (The student should note that numbers should always be represented in this sense, correct to the number of digits given.)

If the numbers to be represented are very large or very small, it is convenient to write them in *floating-point* notation (e.g. the speed of light $2·99793 \times 10^8$ m/s, or the electronic charge $1·6021 \times 10^{-19}$ coulomb). As indicated, we separate the significant figures (the *mantissa*) from the power of ten (the *exponent*); the form in which the exponent is chosen so that the magnitude of the mantissa is less than 10 but not less than 1 is referred to as *scientific notation*.

2 ROUND-OFF ERROR

The simplest way of reducing the number of significant figures in the representation of a number is merely to ignore the unwanted digits.

This procedure, known as *chopping*, is used by many modern computers. A better procedure is *rounding*, which involves adding 5 to the first unwanted digit, and then chopping. For example, π chopped to four decimal places (4*D*) is 3·1415, but it is 3·1416 when rounded; the representation 3·1416 is correct to five significant figures (5*S*). The error involved in the reduction of the number of digits is called *round-off error*. Since π is 3·14159..., we could remark that chopping has introduced much more round-off error than rounding.

3 TRUNCATION ERROR

Numerical results are often obtained by truncating an infinite series or iterative process (see Step 5). Whereas round-off error can be reduced by working to more significant figures, truncation error can be reduced by retaining more terms in the series or more steps in the iteration; this, of course, involves extra work (and perhaps expense!).

4 MISTAKES

In the language of Numerical Analysis, a mistake (or blunder) is *not* an error! A mistake is due to fallibility (usually human, not machine). Mistakes may be trivial, with little or no effect on the accuracy of the calculation, or they may be so serious as to render the calculated results quite wrong. There are three things which may help to eliminate mistakes:
i)   care;
ii)   checks, avoiding repetition;
iii) knowledge of the common sources of mistakes.
Common mistakes include: transposing digits (e.g. reading 6238 as 6328); misreading repeated digits (e.g. reading 62238 as 62338); misreading tables (e.g. referring to a wrong line or a wrong column); incorrectly positioning a decimal point; overlooking signs (especially near sign changes).

5 EXAMPLES

The following illustrate rounding to four decimal places:
$$4/3 \rightarrow 1\cdot3333; \quad \pi/2 \rightarrow 1\cdot5708; \quad 1/\sqrt{2} \rightarrow 0\cdot7071.$$
The following illustrate rounding to four significant figures:
$$4/3 \rightarrow 1\cdot333; \quad \pi/2 \rightarrow 1\cdot571; \quad 1/\sqrt{2} \rightarrow 0\cdot7071.$$

Checkpoint

---

1. What may limit the accuracy of a number in a calculation?
2. What is the convention adopted in rounding?
3. Which would you expect to give better results, rounding or chopping?
4. How can mistakes be avoided?

EXERCISES

1. What are the floating point representations of the following numbers:
   12·345;   0·80059;   296·844; 0·00519?

2. (a) Chop to three significant figures (3S),
   (b) chop to three decimal places (3D),
   (c) round to three significant figures (3S),
   (d) round to three decimal places (3D),
   each of the following numbers:
   34·78219;   3·478219;   0·3478219;   0·03478219.

ERRORS 3
Error propagation and generation

We have noted that a number is to be represented by a finite number of digits, and hence often by an approximation. It is to be expected that the result of any arithmetic procedure (any *algorithm*) involving a set of numbers will have an implicit error relating to the error of the original numbers. We say that the initial errors *propagate* through the computation. In addition, errors may be *generated* at each step in the algorithm, and we may speak of the total cumulative error at any step as the *accumulated* error.

Since we wish to produce results within some chosen limit of error, it is useful to consider error propagation. Roughly speaking from experience, the propagated error depends on the mathematical algorithm chosen, whereas the generated error is more sensitive to the actual ordering of the computational steps. It is possible to be more precise, as is described below.

1 ABSOLUTE ERROR

The *absolute error* is the absolute difference between the exact number x and the approximate number x^*;

i.e. $\qquad e_{abs} = |x - x^*|$.

A number correct to n decimal places has

$$e_{abs} \leqslant 0.5 \times 10^{-n};$$

we expect that the absolute error involved in any approximate number is no more than five units at the first neglected digit.

2 RELATIVE ERROR

The *relative error* is the ratio of the absolute error to the absolute exact number;

i.e. $\qquad e_{rel} = \dfrac{e_{abs}}{|x|} \leqslant \dfrac{e_{abs}}{|x^*| - e_{abs}}$.

(Note that the upper bound follows from the triangle inequality; thus

$$|x^*| = |x + x^* - x| \leqslant |x| + |x^* - x|,$$

so that
$$|x| \geqslant |x^*| - e_{abs} .)$$

If $e_{abs} << |x^*|$, $\text{Max}\{e_{rel}\} \approx \dfrac{e_{abs}}{|x^*|}$.

A decimal number correct to n significant figures has
$$e_{rel} \leqslant 5 \times 10^{-n}.$$

3 ERROR PROPAGATION

Consider two numbers $x = x^* + e_1$, $y = y^* + e_2$.

i)   Under the operations *addition* or *subtraction*,

$$x \mp y = x^* \mp y^* + e_1 \mp e_2$$

so that
$$e \equiv (x \mp y) - (x^* \mp y^*) = e_1 \mp e_2,$$

hence
$$|e| \leqslant |e_1| + |e_2| \; ;$$

i.e.
$$\text{Max}\{|e|\} = |e_1| + |e_2|.$$

The magnitude of the propagated error is therefore not more than the sum of the initial absolute errors; of course, it may be zero.

ii)   Under the operation *multiplication*,

$$xy - x^*y^* = x^*e_2 + y^*e_1 + e_1 e_2,$$

so that
$$\left| \frac{xy - x^*y^*}{x^*y^*} \right| \leqslant \left| \frac{e_1}{x^*} \right| + \left| \frac{e_2}{y^*} \right| + \left| \frac{e_1}{x^*} \frac{e_2}{y^*} \right|$$

and so
$$\text{Max}\{e_{rel}\} \approx \left| \frac{e_1}{x^*} \right| + \left| \frac{e_2}{y^*} \right|,$$

assuming $\left| \dfrac{e_1}{x^*} \dfrac{e_2}{y^*} \right|$ is negligible.

The maximum relative error propagated is approximately the sum of the initial relative errors. The same result is obtained from *division*.

4 ERROR GENERATION

Often (e.g. in a machine) an operation $\otimes$ is also approximated, by an

operation $\otimes^*$, say. Consequently, $x \otimes y$ is represented by $x^* \otimes^* y^*$.
Indeed, one has

$$|x \otimes y - x^* \otimes^* y^*| = |(x \otimes y - x^* \otimes y^*) + (x^* \otimes y^* - x^* \otimes^* y^*)|$$

$$\leqslant |x \otimes y - x^* \otimes y^*| + |x^* \otimes y^* - x^* \otimes^* y^*|,$$

so that the accumulated error does not exceed the sum of the propagated and generated errors. Examples may be found in Step 4.

5 EXAMPLE

Evaluate (as accurately as possible):
(i) $3.45 + 4.87 - 5.16$
(ii) 3.55×2.73

There are two methods which the student may consider, the first of which is to invoke the concepts of absolute and relative error as defined in this Step. Thus the result for (i) is 3.16 ± 0.015, since the maximum absolute error is $0.005 + 0.005 + 0.005 = 0.015$. One concludes that the answer is 3 (to 1S), for the number certainly lies between 3.145 and 3.175. In (ii), the product 9.6915 is subject to the maximum relative error

$$\frac{0.005}{3.55} + \frac{0.005}{2.73} + \frac{0.005}{3.55} \times \frac{0.005}{2.73} \approx (\frac{1}{3.55} + \frac{1}{2.73}) \times 0.005 ;$$

hence maximum (absolute) error $\approx (2.73 + 3.55) \times 0.005 \approx 0.03$, so that the answer is 9.7.

A second approach is to use 'interval arithmetic'. Thus, the approximate number 3.45 represents a number in the interval $(3.445, 3.455)$, etc. Consequently, the result for (i) lies in the interval bounded below by

$$3.445 + 4.865 - 5.165 = 3.145$$

and above by

$$3.455 + 4.875 - 5.155 = 3.175.$$

Similarly, in (ii) the result lies in the interval bounded below by

$$3.545 \times 2.725 \approx 9.66$$

and above by

$$3.555 \times 2.735 \approx 9.72 .$$

Hence one again concludes that the approximate numbers 3 and 9.7 correctly represent the respective results to (i) and (ii).

Checkpoint

1. What distinguishes propagated and generated error?
2. How may the propagated error for the operations addition (subtraction) and multiplication (division) be determined?

EXERCISES

Evaluate the following as accurately as possible, assuming all values are correct to the number of figures given:

(a) $8 \cdot 24 + 5 \cdot 33$;
(b) $124 \cdot 53 - 124 \cdot 52$;
(c) $4 \cdot 27 \times 3 \cdot 13$;
(d) $9 \cdot 48 \times 0 \cdot 513 - 6 \cdot 72$;
(e) $0 \cdot 25 \times 2 \cdot 84 / 0 \cdot 64$;
(f) $1 \cdot 73 - 2 \cdot 16 + 0 \cdot 08 + 1 \cdot 00 - 2 \cdot 23 - 0 \cdot 97 + 3 \cdot 02$.

ERRORS 4
Floating point arithmetic

In Step 2, floating point representation was introduced as a convenient way of dealing with large or small numbers. Since most scientific computation involves such numbers, many students will be familiar with floating point arithmetic and will appreciate the way in which it facilitates calculations involving multiplication or division.

To investigate the implications of finite number representation we need to examine the way in which arithmetic is carried out with floating point numbers. The specifications below apply to most computers that round, and are easily adapted to those that chop. For simplicity in our examples, we will use a 3 digit decimal mantissa, *normalized* to lie in the range $(1, 10)$, i.e. $1 \leqslant |\text{mantissa}| < 10$ (most digital computers use binary representation and the mantissa is commonly normalized to lie in the range $(\frac{1}{2}, 1)$). Note that up to 6 digits are used for intermediate results but the final result of *each* operation is a normalized 3 digit decimal floating point number.

1 ADDITION AND SUBTRACTION

The mantissae are added or subtracted (after shifting the mantissa and increasing the exponent of the smaller number, if necessary, to make the exponents agree); the final result is obtained by rounding (after shifting the mantissa and adjusting the exponent, if necessary). Thus:

$$3{\cdot}12 \times 10^1 \; + \; 4{\cdot}26 \times 10^1 \;\; = \; 7{\cdot}38 \times 10^1$$

$$2{\cdot}77 \times 10^2 \; + \; 7{\cdot}55 \times 10^2 \;\; = \; 10{\cdot}32 \times 10^2 \to 1{\cdot}03 \times 10^3$$

$$6{\cdot}18 \times 10^1 \; + \; 1{\cdot}84 \times 10^{-1} = \; 6{\cdot}18 \times 10^1 + 0{\cdot}0184 \times 10^1$$
$$= \; 6{\cdot}1984 \times 10^1 \to 6{\cdot}20 \times 10^1$$

$$3{\cdot}65 \times 10^{-1} - 2{\cdot}78 \times 10^{-1} = \; 0{\cdot}87 \times 10^{-1} \to 8{\cdot}70 \times 10^{-2}$$

2 MULTIPLICATION

The exponents are added and the mantissae are multiplied; the final result is obtained by rounding (after shifting the mantissa right and increasing the exponent by 1, if necessary). Thus:

$$(4\cdot27 \times 10^1) \times (3\cdot68 \times 10^1) = 15\cdot7136 \times 10^2 \to 1\cdot57 \times 10^3$$

$$(2\cdot73 \times 10^2) \times (-3\cdot64 \times 10^{-2}) = -9\cdot9372 \times 10^0 \to -9\cdot94 \times 10^0.$$

3 DIVISION

The exponents are subtracted and the mantissae are divided; the final result is obtained by rounding (after shifting the mantissa left and reducing the exponent by 1, if necessary). Thus:

$$5\cdot43 \times 10^1/(4\cdot55 \times 10^2) = 1\cdot19340\ldots \times 10^{-1} \to 1\cdot19 \times 10^{-1}$$

$$-2\cdot75 \times 10^2/(9\cdot87 \times 10^{-2}) = -0\cdot278622\ldots \times 10^4 \to -2\cdot79 \times 10^3$$

4 EXPRESSIONS

The order of evaluation is determined in a standard way and the result of *each* operation is a normalized floating point number. Thus:

$$(6\cdot18 \times 10^1 + 1\cdot84 \times 10^{-1})/((4\cdot27 \times 10^1) \times (3\cdot68 \times 10^1))$$

$$\to 6\cdot20 \times 10^1/(1\cdot57 \times 10^3) = 3\cdot94904\ldots \times 10^{-2} \to 3\cdot95 \times 10^{-2}$$

5 GENERATED ERROR

We note that all the above examples (except the subtraction and the first addition) involve generated errors which are relatively large because of the small size of the mantissae. Thus the generated error in

$$2\cdot77 \times 10^2 + 7\cdot55 \times 10^2 = 10\cdot32 \times 10^2 \to 1\cdot03 \times 10^3$$

is $0\cdot002 \times 10^3$. Since the propagated error in this example may be as large as $0\cdot01 \times 10^2$ (assuming the operands are correct to $3S$), we can use the result of Step 3 to deduce that the accumulated error cannot exceed $0\cdot002 \times 10^3 + 0\cdot01 \times 10^2 = 0\cdot003 \times 10^3$.

6 CONSEQUENCES

The pecularities of floating point arithmetic lead to some unexpected and unfortunate consequences, including the following:
a)   Addition or subtraction of a small (but non-zero) number may have no effect, e.g.

$$5 \cdot 18 \times 10^2 + 4 \cdot 37 \times 10^{-1} = 5 \cdot 18 \times 10^2 + 0 \cdot 00437 \times 10^2$$
$$= 5 \cdot 18437 \times 10^2 \to 5 \cdot 18 \times 10^2,$$

(i.e. the additive identity is non-unique).

b) Frequently the result of $a \times (1/a)$ is not 1,

   e.g., if $a = 3 \cdot 00 \times 10^0$ then $1/a \to 3 \cdot 33 \times 10^{-1}$

   and $a \times (1/a) \to 9 \cdot 99 \times 10^{-1}$,

   (i.e. the multiplicative inverse may not exist).

c) The result of $(a + b) + c$ is not always the same as the result of $a + (b + c)$,

   e.g. $a = 6 \cdot 31 \times 10^1$, $b = 4 \cdot 24 \times 10^0$, $c = 2 \cdot 47 \times 10^{-1}$,

   then $(a + b) + c = (6 \cdot 31 \times 10^1 + 0 \cdot 424 \times 10^1) + 2 \cdot 47 \times 10^{-1}$

   $$\to 6 \cdot 73 \times 10^1 + 0 \cdot 0247 \times 10^1$$

   $$\to 6 \cdot 75 \times 10^1,$$

   whereas $a + (b + c) = 6 \cdot 31 \times 10^1 + (4 \cdot 24 \times 10^0 + 0 \cdot 247 \times 10^0)$

   $$\to 6 \cdot 31 \times 10^1 + 4 \cdot 49 \times 10^0$$

   $$\to 6 \cdot 31 \times 10^1 + 0 \cdot 449 \times 10^1$$

   $$\to 6 \cdot 76 \times 10^1,$$

(i.e. the associative law for addition does not always hold).

Examples involving adding many numbers of varying size indicate that adding in order of increasing magnitude is preferable to adding in the reverse order.

Checkpoint

---

1. Why is it sometimes necessary to shift the mantissa and adjust the exponent of a floating point number?
2. Does floating point arithmetic obey the usual laws of arithmetic?

EXERCISES

1. Evaluate the following using 3 digit decimal normalized floating point arithmetic with rounding:
   a) $6 \cdot 19 \times 10^2 + 5 \cdot 82 \times 10^2$;
   b) $6 \cdot 19 \times 10^2 + 3 \cdot 61 \times 10^1$;
   c) $6 \cdot 19 \times 10^2 - 5 \cdot 82 \times 10^2$;
   d) $6 \cdot 19 \times 10^2 - 3 \cdot 61 \times 10^1$;
   e) $(3 \cdot 60 \times 10^3) \times (1 \cdot 01 \times 10^{-1})$;
   f) $(-7 \cdot 50 \times 10^{-1}) \times (-4 \cdot 44 \times 10^1)$
   g) $(6 \cdot 45 \times 10^2) / (5 \cdot 16 \times 10^{-1})$;
   h) $(-2 \cdot 86 \times 10^{-2}) / (3 \cdot 29 \times 10^3)$.

2. Estimate the accumulated errors in the results of Exercise 1, assuming that all values are correct to $3S$.

3. Evaluate the following, using 4 digit decimal normalized floating point arithmetic with rounding, then recalculate carrying all decimal places and estimate the propagated error.

   a) Given $a = 6 \cdot 842 \times 10^{-1}$, $b = 5 \cdot 685 \times 10^1$, $c = 5 \cdot 641 \times 10^1$, find $a(b - c)$ and $ab - ac$.

   b) Given $a = 9 \cdot 812 \times 10^1$, $b = 4 \cdot 631 \times 10^{-1}$, $c = 8 \cdot 340 \times 10^{-1}$, find $(a + b) + c$ and $a + (b + c)$.

ERRORS 5
Approximation to functions

An important procedure in Analysis is to represent a given function as an infinite series of terms involving simpler or otherwise more appropriate functions. Thus, if $f(x)$ is the given function, it may be represented as the *series expansion*

$$f(x) = a_0\phi_0(x) + a_1\phi_1(x) + \ldots + a_n\phi_n(x) + \ldots,$$

involving the set of functions $\{\phi_j(x)\}$. Mathematicians have spent a lot of effort in discussing the *convergence* of series; i.e. in defining conditions for which the partial sum

$$s_n(x) = a_0\phi_0(x) + a_1\phi_1(x) + \ldots + a_n\phi_n(x)$$

approximates the function value $f(x)$ ever more closely as n increases. In Numerical Analysis, we are primarily concerned with such convergent series; computing the sequence of partial sums is an approximation process in which the *truncation* error may be made as small as we please by taking sufficient terms into account.

1 THE TAYLOR SERIES

The most important expansion to represent a function is the *Taylor series*. If $f(x)$ is suitably smooth in the neighbourhood of some chosen point x_0 we have

$$f(x) = f(x_0) + hf'(x_0) + \frac{1}{2!}h^2f''(x_0) + \ldots + \frac{1}{n!}h^nf^{(n)}(x_0) + R_n,$$

where $f^{(k)}(x_0) \equiv \dfrac{d^kf}{dx^k}\bigg|_{x=x_0};$

$$h = x - x_0$$

denotes the displacement from x_0 to point x in the neighbourhood, and the remainder term is

$$R_n = \frac{1}{(n+1)!}\, h^{n+1} f^{(n+1)}\, (\xi)$$

for some point ξ between x_0 and x. (This is known as the Lagrange form of the remainder; see for example Section 18·4 in G.B. Thomas's book cited in the Bibliography.)

The Taylor expansion converges for x within some range including the point x_0, a range which lies within the neighbourhood of x_0 mentioned above. Within this *range of convergence*, the *truncation error* due to discarding terms after the nth (equal to the value of R_n at point x) can be made smaller in magnitude than any positive constant by choosing n sufficiently large. In other words, by using R_n to decide how many terms are needed, one may evaluate the function at any point in the range of convergence as accurately as the accumulation of round-off error permits.

From the viewpoint of the numerical analyst, it is most important that the *convergence is fast enough*. For example, if we set $f(x) = \sin x$ we have

$$f'(x) = \cos x\,,$$

$$f''(x) = -\sin x\,,$$

etc.,

and the expansion (about $x_0 = 0$)

$$\sin x = x - \frac{x^3}{3!} + \frac{x^5}{5!} - \ldots + \frac{(-1)^{n-1}x^{2n-1}}{(2n-1)!} + R_n$$

with

$$R_n = \frac{(-1)^n x^{2n+1}}{(2n+1)!}\ \cos\xi.$$

Since $|\cos \xi| \leqslant 1, |R_n| \leqslant \dfrac{|x|^{2n+1}}{(2n+1)!};$ if $5D$ accuracy is required, it

follows that only 2 terms are needed at $x = 0\cdot1$, and 4 terms at $x = 1$ (since $9! = 362,880$). On the other hand, the expansion for the natural (base e) logarithm,

$$\log_e(1+x) = x - \frac{x^2}{2} + \frac{x^3}{3} - \ldots + \frac{(-1)^{n-1}x^n}{n} + R_n$$

is less suitable. Although only 4 terms are needed to give $5D$ accuracy at $x = 0\cdot1$, 13 terms are required for $5D$ accuracy at $x = 0\cdot5$, and 10 terms give just $1D$ accuracy at $x = 1$!

Further, we remark that the Taylor series is not only used extensively to represent functions numerically, but also to analyse the errors involved in various algorithms (e.g. see Steps 8, 9, 10, 27 and 28).

2 POLYNOMIAL APPROXIMATION

The Taylor series provides a simple method of *polynomial approximation* (of chosen degree n),

$$f(x) \approx a_0 + a_1 x + a_2 x^2 + \ldots + a_n x^n,$$

which is basic to the discussion of various elementary numerical procedures in this textbook. Because $f(x)$ is often complicated, one may prefer to execute operations such as differentiation and integration on a polynomial approximation. Interpolation formulae (see Steps 20 and 22) may also be used to construct polynomial approximations.

3 OTHER SERIES EXPANSIONS

There are many other series expansions, such as the *Fourier series* (in terms of sines and cosines), or those involving various *orthogonal functions* (Legendre polynomials, Chebyshev polynomials, Bessel functions, etc.) From the numerical standpoint, truncated Fourier series and Chebyshev polynomial series have proven to be the most useful. Fourier series are appropriate in dealing with functions with natural periodicity, while Chebyshev series provide the most rapid convergence of all known approximations based on polynomials.

Occasionally, it is possible to represent a function adequately (from the numerical standpoint) by truncating a series which does *not* converge in the mathematical sense. For example, solutions are sometimes obtained in the form of *asymptotic series* with leading terms which provide sufficiently accurate numerical results.

While we confine our attention in this book to truncated Taylor series, the interested student should be aware that such alternative expansions exist (see for example Conte and de Boor).

4 RECURSIVE PROCEDURES

While a truncated series with few terms may be a practical way to compute values of a function, there is a number of arithmetic operations involved. Particularly in automatic computation, some available *recursive procedure* which reduces the amount of arithmetic may be favoured. For example, the values of the polynomial

$$P(x) = a_0 + a_1 x + a_2 x^2 + \ldots\ldots + a_n x^n,$$

and its derivative

$$P'(x) = a_1 + 2a_2 x + \ldots\ldots + na_n x^{n-1},$$

for $x = \bar{x}$ may be generated recursively under the scheme:

$$p_0 = a_n, q_0 = 0$$

$$p_k = p_{k-1}\bar{x} + a_{n-k}$$
$$\qquad\qquad\qquad\qquad k = 1, 2, \ldots\ldots, n.$$
$$q_k = q_{k-1}\bar{x} + p_{k-1}$$

Thus, for successive values of k one has

$$p_1 = p_0\bar{x} + a_{n-1} = a_n\bar{x} + a_{n-1}, \qquad\qquad q_1 = q_0\bar{x} + p_0 = a_n ;$$

$$p_2 = p_1\bar{x} + a_{n-2} = a_n\bar{x}^2 + a_{n-1}\bar{x} + a_{n-2}, \quad q_2 = q_1\bar{x} + p_1 = 2a_n\bar{x}$$
$$+ a_{n-1} ;$$

$$\qquad \cdot \qquad\qquad \cdot \qquad\qquad\qquad\qquad\qquad\qquad \cdot \qquad\qquad \cdot$$
$$\qquad \cdot \qquad\qquad \cdot \qquad\qquad\qquad\qquad\qquad\qquad \cdot \qquad\qquad \cdot$$

$$P_n = P(\bar{x}), \qquad\qquad\qquad\qquad\qquad\qquad q_n = P'(\bar{x}).$$

(Perhaps the student may be able to suggest a recursive procedure for even higher derivatives of $P(x)$.)

Finally, it should be noted that it is common to generate members of a set of orthogonal functions recursively.

Checkpoint

---

1. How do numerical analysts use the remainder term R_n in Taylor series?
2. Why is 'speed of convergence' so important from the numerical standpoint?
3. From the numerical standpoint, is it essential for a series representation to converge in the mathematical sense?

EXERCISES

1. Use the Taylor series to expand $\cos x$ about $x = 0$.
2. Truncate the Taylor series to give linear, quadratic and cubic polynomial

approximations for $f(x) = e^x$ in the neighbourhood of $x = 0$. Use the remainder term to estimate (to the nearest 0·1) the range over which each polynomial approximation yields results correct to 2D.

3. Find the number of terms required in the Taylor series for $f(x) = e^x$ about $x = 0$ to give 5D accuracy for all x between 0 and 1.

4. Evaluate $P(3·1)$ and $P'(3·1)$ where $P(x) = x^3 - 2x^2 + 2x + 3$.

NON-LINEAR EQUATIONS 1
Solving non-linear algebraic and transcendental equations

The first non-linear equation encountered in algebra courses is usually the quadratic equation.

$$ax^2 + bx + c = 0,$$

and all students will be familiar with the formula for its roots:

$$x = \frac{-b \pm \sqrt{b^2 - 4ac}}{2a}.$$

The formula for the roots of a general cubic is somewhat more complicated and that for a general quartic usually takes several pages to describe! We are spared further effort by a theorem which states that there is no such formula for general polynomials of degree higher than four. Accordingly, except in special cases (e.g. when factorization is easy), we prefer in practice to use a *numerical* method to solve polynomial equations of degree higher than two.

Another class of non-linear equations consists of those which involve *transcendental* functions such as e^x, log x, sin x and tan x. Useful analytic solutions of such equations are rare so we are usually forced to use numerical methods.

1 A TRANSCENDENTAL EQUATION

We shall use a simple mathematical problem to show that transcendental equations do arise quite naturally. Suppose we seek the height of liquid in a cylindrical tank of radius r, lying with its axis horizontal, when the tank is a quarter full (see Figure 2). Suppose the height of liquid is h (DB in the diagram). The condition to be satisfied is that the area of the *segment* ABC should be $\frac{1}{4}$ of the area of the circle. This reduces to

$$2 \left[\tfrac{1}{2} r^2\theta - \tfrac{1}{2}(r \sin \theta)(r \cos \theta) \right] = \tfrac{1}{4} \pi r^2.$$

($\tfrac{1}{2} r^2\theta$ is the area of the *sector* OAB, $r \sin \theta$ is the base and $r \cos \theta$ the height of the *triangle* OAD.)

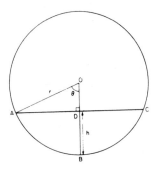

FIGURE 2. Cylindrical tank (Cross-Section)

Hence $2\theta - 2\sin\theta\cos\theta = \dfrac{\pi}{2}$

or $x + \cos x = 0$,   where $x = \dfrac{\pi}{2} - 2\theta$

(since $2\sin\theta\cos\theta = \sin 2\theta = \sin\left(\dfrac{\pi}{2} - x\right) = \cos x$).

When we have solved the transcendental equation

$$f(x) \equiv x + \cos x = 0,$$

we obtain h from

$$h = OB - OD$$
$$= r - r\cos\theta$$
$$= r\left(1 - \cos\left(\dfrac{\pi}{4} - \dfrac{x}{2}\right)\right).$$

2 LOCATING ROOTS

Let us suppose that our problem is to find some or all of the roots of
the non-linear equation $f(x) = 0$. Before we use a numerical method
(c.f. Steps 7-10) we should have some idea about the number, nature
and approximate location of the roots. The usual approach involves
the construction of *graphs* or *tables of values* of the function $f(x)$, and
is best illustrated by examples.

a) $\sin x - x + 0{\cdot}5 = 0$

As often happens, it is easier to separate $f(x)$ into two parts, sketch
two curves on the one set of axes, and see where they intersect.

In this case we sketch $y = \sin x$ and $y = x - 0{\cdot}5$. Since $|\sin x| \leqslant 1$

we are only interested in the interval $-0.5 \leqslant x \leqslant 1.5$ (outside which $|x - 0.5| > 1$).

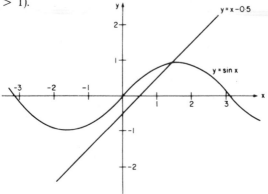

FIGURE 3.   Graphs of $y = x - 0.5$ and $y = \sin x$

We deduce from the graph that the equation has only one real root, near $x = 1.5$. We then tabulate $f(x) \equiv \sin x - x + 0.5$ near $x = 1.5$ as follows:

x	1·5	1·45	1·49
$\sin x$	0·9975	0·9927	0·9967
$f(x)$	−0·0025	0·0427	0·0067

We now know that the root lies between 1·49 and 1·50, and we can use a numerical method to obtain a more accurate answer.

b) $e^{-0.2x} = x(x-2)(x-3)$.

Again we sketch two curves:

$$y = e^{-0.2x}$$

and

$$y = x(x-2)(x-3).$$

In sketching the second curve we use the three obvious zeros at $x = 0$, 2 and 3, as well as the knowledge that $x(x-2)(x-3)$ is negative for $x < 0$ and $2 < x < 3$ and positive and increasing steadily for $x > 3$. We deduce from the graph that there are three real roots, near $x = 0.2$,

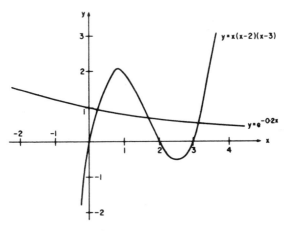

FIGURE 4. Graphs of y

1·8 and 3·1, and tabulate as follows
(with $f(x) = e^{-0.2x} - x(x - 2)(x - 3)$):

x	0·2	0·15	1·8	1·6	3·1	3·2
$e^{-0.2x}$	0·9608	0·9704	0·6977	0·7261	0·5379	0·5273
$x(x-2)(x-3)$	1·0080	0·7909	0·4320	0·8960	0·3410	0·7680
$f(x)$	−0·0472	0·1795	0·2657	−0·1699	0·1969	−0·2407

We conclude that the roots lie between 0·15 and 0·2, 1·6 and 1·8, and
3·1 and 3·2, respectively.

Checkpoint

1. Why are numerical methods used in solving non-linear
    equations?
2. How does a transcendental equation differ from an algeb-
    raic equation?
3. What kind of information is used in sketching curves for
    root location?

EXERCISE

Locate the roots of the equation

$$x + \cos x = 0.$$

NON-LINEAR EQUATIONS 2
The bisection method

The *bisection method*[†] for solving the equation $f(x) = 0$ for the values of x (the *roots*) is based on the following theorem.

THEOREM: If $f(x)$ is continuous for x between a and b and if $f(a)$ and $f(b)$ have opposite signs, then there exists at least one real root of $f(x) = 0$ between a and b.

1 PROCEDURE

Suppose that a continuous function $f(x)$ is negative at $x = a$ and positive at $x = b$, so that there is at least one real root between a and b. (Usually a and b are found by curve sketching.) If we calculate $f[(a + b)/2]$, which is the function value at the point of bisection of the interval $a < x < b$, there are three possibilities:

i) $f[(a + b)/2] = 0$ in which case $(a + b)/2$ is the root;

ii) $f[(a + b)/2] < 0$ in which case the root lies between $\dfrac{a+b}{2}$ and b;

iii) $f[(a + b)/2] > 0$ in which case the root lies between $\dfrac{a+b}{2}$ and a.

   Presuming there is just one root, if case (i) occurs the process is terminated. If either case (ii) or case (iii) occurs, the process of bisection of the interval containing the root can be repeated until the root is obtained to the desired accuracy. In Figure 5, the successive points of bisection are denoted by x_1, x_2 and x_3.

2 EFFECTIVENESS

The bisection method is suitable for automatic computation and is almost certain to give a root. Provided the conditions of the above theorem hold, it can only fail if the accumulated error in the calculation

† This algorithm is suitable for automatic computation. A flow-chart for study and use in programming may be found on page 150.

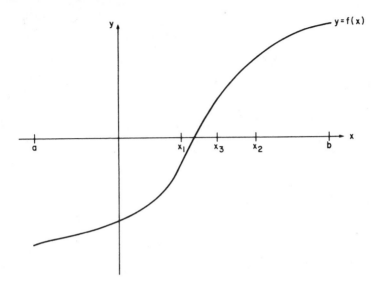

FIGURE 5. Successive bisection

of $f(x)$ at a bisection point gives it a small negative value when actually it should have a small positive value (or vice versa); the interval subsequently chosen would therefore be wrong. This can be overcome by working to sufficient accuracy, and this almost-assured convergence is not true of many other methods of finding a root.

One drawback of the bisection method is that it applies only for roots of $f(x)$ about which $f(x)$ changes sign. In particular, double roots can be overlooked; one should be careful to examine $f(x)$ in any range where it is small, so that repeated roots about which $f(x)$ does not change sign are otherwise evaluated (e.g. see Steps 9 and 10). Such a close examination of course also avoids another nearby root being overlooked.

Finally, note that bisection is rather slow; after n steps the interval containing the root is of length $|b - a|/2^n$. However, provided values of $f(x)$ can be generated readily, the rather large number of steps which can be involved in the application of bisection is of relatively little consequence for an *automatic* computer!

3 EXAMPLE

Solve $3xe^x = 1$ to 3 decimal places by the bisection method. We can consider $f(x) = 3x - e^{-x}$, which changes sign in the interval $0.25 < x < 0.27$: one may tabulate (working to 4 D) as follows:

x	$3x$	e^{-x}	$f(x)$
0·25	0·75	0·7788	$-$ 0·0288
0·27	0·81	0·7634	$+$ 0·0466

(The student should ascertain that there is just one root.)
Proceeding to bisection:

a	b	$x = \dfrac{a+b}{2}$	$3x$	e^{-x}	$f(x)$
0·25	0·27	0·26	0·78	0·7711	$+$ 0·0089
0·25	0·26	0·255	0·765	0·7749	$-$ 0·0099
0·255	0·26	0·2575	0·7725	0·7730	$-$ 0·0005
0·2575	0·26	0·2588	0·7764	0·7720	$+$ 0·0044
0·2575	0·2588	0·2582	0·7746	0·7724	$+$ 0·0022

Hence $0·2575 < x < 0·2582$, so that the root is $0·258$.

Checkpoint

1. When may the bisection method be used to find a root of the equation $f(x) = 0$?
2. What are the three possible choices after a bisection value is calculated?
3. What is the maximum error after n bisections?

EXERCISES

1. Use the bisection method to find the root of the equation

$$x + \cos x = 0$$

correct to two decimal places (2D).
2. Use the bisection method to find the positive root of the equation

$$x - 0·2 \sin x - 0·5 = 0$$

to 3D.

NON-LINEAR EQUATIONS 3
Method of false position

As mentioned in the *Prologue*, the *method of false position*[†] dates back to the ancient Egyptians. It remains an effective alternative to the bisection method for solving the equation $f(x) = 0$ for a real root between a and b, given that $f(x)$ is continuous and $f(a)$ and $f(b)$ have opposite signs.

1 PROCEDURE

The curve $y = f(x)$ is not generally a straight line. However, one may join the points

$$(a, f(a)) \quad \text{and} \quad (b, f(b))$$

by the straight line

$$\frac{y - f(a)}{f(b) - f(a)} = \frac{x - a}{b - a}$$

The straight line cuts the x-axis at $(\bar{x}, 0)$ where

$$\frac{0 - f(a)}{f(b) - f(a)} = \frac{\bar{x} - a}{b - a}$$

so that

$$\bar{x} = a - f(a) \frac{b - a}{f(b) - f(a)}$$

$$= \frac{af(b) - bf(a)}{f(b) - f(a)} = \frac{1}{f(b) - f(a)} \begin{vmatrix} a & f(a) \\ b & f(b) \end{vmatrix}$$

Let us suppose that $f(a)$ is negative and $f(b)$ is positive. As in the bisection method, there are three possibilities:

i) $f(\bar{x}) = 0$, in which case $\bar{x}$ is the root;

ii) $f(\bar{x}) < 0$, in which case the root lies between $\bar{x}$ and b;

[†] This algorithm is suitable for automatic computation. A flow-chart for study and use in programming may be found on page 151.

iii) $f(\bar{x}) > 0$, in which case the root lies between $\bar{x}$ and a.

Again, if case (i) occurs, the process is terminated; if either case (ii) or case (iii) occurs, the process can be repeated until the root is obtained to the desired accuracy. In Figure 6, the successive points where the straight lines cut the axis are denoted by x_1, x_2, x_3.

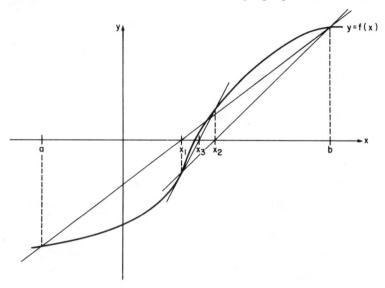

FIGURE 6. Method of false position

2 EFFECTIVENESS AND THE SECANT METHOD

Like the bisection method, the method of false position has almost assured convergence, and it may converge to a root faster. However, it may happen that most or all of the calculated $\bar{x}$ values are on the same side of the root, in which case convergence may be slow (see Figure 7). This is avoided in the *secant method*, which resembles the method of false position except that no attempt is made to ensure that the root is enclosed. Starting with two approximations to the root (x_0 and x_1) further approximations x_2, x_3, ... are computed from

$$x_{n+1} = x_n - f(x_n)\frac{x_n - x_{n-1}}{f(x_n) - f(x_{n-1})}.$$

We no longer have assured convergence, but the process is simpler (the sign of $f(x_{n+1})$ is not tested) and often converges faster.

With respect to speed of convergence of the secant method, we have the error at the $(n+1)$ th step:

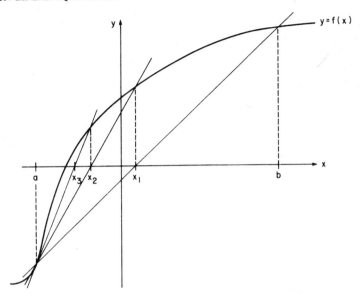

FIGURE 7. Method of false position

$$e_{n+1} = x - x_{n+1}$$

$$= \frac{(x-x_{n-1})f(x_n) - (x-x_n)f(x_{n-1})}{f(x_n) - f(x_{n-1})}$$

$$= \frac{e_{n-1}f(x-e_n) - e_n f(x-e_{n-1})}{f(x-e_n) - f(x-e_{n-1})}.$$

Hence, expanding in terms of the Taylor series,

$$e_{n+1} = \frac{e_{n-1}\left[f(x)-e_n f'(x)+(e_n^2/2)f''(x)-...\right] - e_n\left[f(x)-e_{n-1}f'(x)+(e_{n-1}^2/2)f''(x)-...\right]}{\left[f(x)-e_n f'(x)+...\right] - \left[f(x)-e_{n-1}f'(x)+...\right]}$$

$$\approx \quad \left[\frac{-f''(x)}{2f'(x)}\right] e_{n-1}e_n \sim e_{n-1}e_n.$$

We seek k such that $e_n \sim e_{n-1}^k$; then $e_{n+1} \sim e_n^k \sim e_{n-1}^{k^2}$ and $e_{n-1}e_n \sim e_{n-1}^{k+1}$, so that we deduce $k^2 \approx k+1$, whence $k \approx (1+\sqrt{5})/2 = 1\cdot618$. The speed of convergence is therefore faster than linear ($k=1$), but slower than quadratic ($k=2$).

3 EXAMPLE

Solve $3xe^x = 1$ to 3 decimal places by the method of false position.

In the previous step, we observed that the root lies in the interval $0.25 < x < 0.27$. Consequently,

$$\bar{x} = \frac{1}{0.0466 + 0.0288} \begin{vmatrix} 0.25 & -0.0288 \\ 0.27 & 0.0466 \end{vmatrix}$$

$$= \frac{0.01165 + 0.00778}{0.0754} = 0.2577 .$$

Writing $f(x) = 3x - e^{-x}$ as before,

$$f(\bar{x}) = f(0.2577)$$

$$= 3 \times 0.2577 - 0.7728$$

$$= 0.7731 - 0.7728$$

$$= 0.0003.$$

It is tempting to conclude that the root is 0.258, but it is necessary to confirm this. Thus, we know that the root is in the interval $0.25 < x < 0.2577$, so that repeating the process

$$\bar{x} = \frac{1}{0.0003 + 0.0288} \begin{vmatrix} 0.25 & -0.0288 \\ 0.2577 & 0.0003 \end{vmatrix}$$

$$= \frac{0.000075 + 0.007422}{0.0291} = 0.2576 ,$$

$$f(\bar{x}) = -0.0001.$$

Hence $0.2576 < x < 0.2577$, so that the root is 0.258.

Checkpoint

---

1. When may the method of false position be used to find a root of the equation $f(x) = 0$?
2. On what geometric construction is the method of false position based?

EXERCISES

1. Use the method of false position to find the smallest positive root of the equation $2 \sin x + x - 2 = 0$ correct to three decimal places.

2. Compare the results obtained when the bisection method, the method of false position and the secant method are used (with starting values 0·7 and 0·9) to solve the equation

$$3 \sin x = x + \frac{1}{x}.$$

3. Use the method of false position to find the root of the equation

$$x + \cos x = 0$$

to 4D.

NON-LINEAR EQUATIONS 4
The method of simple iteration

The method of simple iteration involves writing the equation $f(x) = 0$ in a form $x = \phi(x)$, suitable for the construction of a sequence of approximations to some root, in a repetitive fashion.

1 PROCEDURE

The iteration procedure is as follows. In some way we obtain a rough approximation x_0 of the desired root, which may then be substituted into the right hand side to give a new approximation, $x_1 = \phi(x_0)$. The new approximation is again substituted into the right hand side to give a further approximation $x_2 = \phi(x_1)$, and so on until (hopefully) a sufficiently accurate approximation to the root is obtained. This repetitive process, based on $x_{n+1} = \phi(x_n)$, is called *simple iteration*; provided that $|x_{n+1} - x_n|$ decreases as n increases, the process tends to $\alpha = \phi(\alpha)$, where α denotes the root.

2 EXAMPLE

Find the root of the equation $3xe^x = 1$ to an accuracy of $0 \cdot 0001$, using the method of simple iteration.
    One first writes

$$x = \tfrac{1}{3}e^{-x}$$
$$= \phi(x).$$

Assuming $x_0 = 1$, successive iterations produce

$$x_1 = 0 \cdot 12263,$$
$$x_2 = 0 \cdot 29486,$$
$$x_3 = 0 \cdot 24821,$$
$$x_4 = 0 \cdot 26007,$$

$$x_5 = 0.25700,$$

$$x_6 = 0.25779,$$

$$x_7 = 0.25759,$$

$$x_8 = 0.25764.$$

Thus, eight iterations are necessary before $|x_{n+1} - x_n| < 0.0001$. A graphical interpretation of the first three iterations is shown in Figure 8.

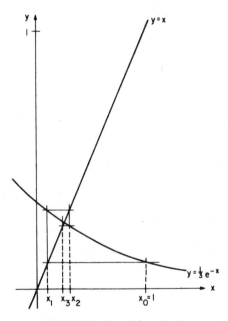

FIGURE 8. Iterative method

3 CONVERGENCE

Whether or not an iteration procedure converges quickly, or indeed at all, depends on the choice of the function $\phi(x)$, as well as the starting value x_0. For example, the equation $x^2 = 3$ has two real roots, $\pm\sqrt{3}$ $(= \pm 1.732)$. It can be rewritten in the form

$$x = \frac{3}{x} \equiv \phi(x),$$

which suggests the iteration

$$x_{n+1} = \frac{3}{x_n}.$$

However, if the starting value $x_0 = 1$ is used, successive iterations give

$$x_1 = \frac{3}{x_0} = 3,$$

$$x_2 = \frac{3}{x_1} = 1,$$

$$x_3 = \frac{3}{x_2} = 3, \quad \text{etc!}$$

We can examine the convergence of the iteration process

$$x_{n+1} = \phi(x_n)$$

to

$$\alpha = \phi(\alpha)$$

with the help of the Taylor series

$$\phi(\alpha) = \phi(x_k) + (\alpha - x_k)\, \phi'(\zeta_k), \quad k = 0, 1, ..., n,$$

where ζ_k is a point between the root α and the approximation x_k. We have

$$\alpha - x_1 = \phi(\alpha) - \phi(x_0) = (\alpha - x_0)\, \phi'(\zeta_0)$$

$$\alpha - x_2 = \phi(\alpha) - \phi(x_1) = (\alpha - x_1)\, \phi'(\zeta_1)$$

$$\vdots \qquad \qquad \vdots$$

$$\alpha - x_{n+1} = \phi(\alpha) - \phi(x_n) = (\alpha - x_n)\, \phi'(\zeta_n).$$

Multiplying the $n+1$ rows together and cancelling the common factors $\alpha - x_1, \alpha - x_2, ..., \alpha - x_n$ leaves

$$\alpha - x_{n+1} = (\alpha - x_0)\, \phi'(\zeta_0) \phi'(\zeta_1) \ ... \ \phi'(\zeta_n).$$

Consequently,

$$|\alpha - x_{n+1}| = |\alpha - x_0|\, |\phi'(\zeta_0)|\, |\phi'(\zeta_1)| \, ... \, |\phi'(\zeta_n)|,$$

so that the absolute error $|\alpha - x_{n+1}|$ can be made as small as we please by sufficient iteration *if* $|\phi'| < 1$ *in the neighbourhood of the root.*

(Note that $\phi(x) = \dfrac{3}{x}$ has derivative $|\phi'(x)| = \left| -\dfrac{3}{x^2} \right| > 1$ for $|x| < \sqrt{3}$.)

Checkpoint

1. What should a programmer guard against in an automatic routine using the method of simple iteration?
2. What is necessary to ensure that the method of simple iteration does converge to a root?

EXERCISES

1. Assuming the initial guess $x_0 = 1$, show by the method of simple iteration that one root of the equation $2x - 1 - 2 \sin x = 0$ is 1.4973.
2. Use the method of simple iteration to find (to $4D$) the root of the equation $x + \cos x = 0$.

NON-LINEAR EQUATIONS 5
The Newton-Raphson iterative method

The *Newton-Raphson method*† is a process for the determination of a root of an equation $f(x) = 0$, given just one point close to the desired root. It can be viewed as a limiting case of the secant method (see Step 8) or as a special case of the method of simple iteration (see Step 9).

1 PROCEDURE

Let x_0 denote the known approximate value of the root of $f(x) = 0$, and let h denote the difference between the true value α and the approximate value;

i.e. $\qquad \alpha = x_0 + h.$

The second degree terminated Taylor expansion about x_0 is

$$f(\alpha) = f(x_0 + h) = f(x_0) + hf'(x_0) + \frac{h^2}{2}f''(x_0 + \theta h), \qquad 0 < \theta < 1.$$

Ignoring the remainder term and writing $f(\alpha) = 0$,
$$f(x_0) + hf'(x_0) \approx 0$$

so $\qquad h \approx -\dfrac{f(x_0)}{f'(x_0)},$

and consequently

$$x_1 = x_0 - \frac{f(x)}{f'(x_0)}$$

should be a better estimate of the root than x_0.

Even better approximations may be obtained by repetition (iteration) of the process, which may then be written as

$$x_{n+1} = x_n - \frac{f(x_n)}{f'(x_n)}.$$

† This algorithm is suitable for automatic computation. A flow-chart for study and use in programming may be found on page 152.

Note that if $f(x)$ is a polynomial we use the recursive procedure of Step 5 to compute $f(x_n)$ and $f'(x_n)$.

The geometrical interpretation is that each step provides the point at which the tangent at the original point cuts the x-axis (c.f. Figure 9).

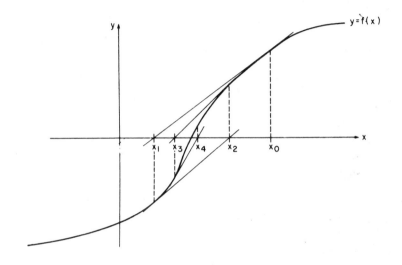

FIGURE 9.   Newton-Raphson Method

Thus, the equation of the tangent at $(x_0, f(x_0))$ is

$$y - f(x_0) = f'(x_0)(x - x_0),$$

so that $(x_1, 0)$ corresponds to

$$- f(x_0) = f'(x_0)(x_1 - x_0)$$

or
$$x_1 = x_0 - \frac{f(x_0)}{f'(x_0)}$$

2   EXAMPLE

Find the positive root of the equation $\sin x = x^2$ correct to 3 decimals, using the Newton-Raphson method.

It is convenient to use the method of false position to obtain an initial approximation. Tabulating, one has

x	$f(x) = \sin x - x^2$
0	0
0·25	0·18
0·5	0·23
0·75	0·12
1	−0·16

There is a root in the interval $0.75 < x < 1$ at approximately

$$x_0 = \frac{1}{-0.16 - 0.12} \begin{vmatrix} 0.75 & 0.12 \\ 1 & -0.16 \end{vmatrix}$$

$$= -\frac{1}{0.28}(-0.12 - 0.12)$$

$$= \frac{0.24}{0.28} = 0.857$$

We now use the Newton-Raphson method; we have

$$f(0.857) = \sin(0.857) - (0.857)^2$$

$$= 0.7559 - 0.7344$$

$$= 0.0215$$

and $\qquad f'(x) \quad = \cos x - 2x,$

giving $\qquad f'(0.857) = 0.6547 - 1.714$

$$= -1.0593.$$

Consequently, a better approximation is

$$x_1 = 0.857 + \frac{0.0215}{1.0593}$$

$$= 0.857 + 0.0203$$

$$= 0.8773$$

Repeating the procedure, we obtain

$$f(x_1) \quad = f(0.8773) = -0.0007$$

$$f'(x_1) \quad = f'(0.8773) = -1.1154$$

so that

$$x_2 = 0.8773 - \frac{0.0007}{1.1154}$$

$$= 0.8773 - 0.0006$$

$$= 0.8767$$

Since $f(x_2) = 0.0000$, we conclude that the root is 0.877 to $3D$.

3 CONVERGENCE

If we write

$$\phi(x) = x - \frac{f(x)}{f'(x)},$$

the Newton-Raphson iteration expression

$$x_{n+1} = x_n - \frac{f(x_n)}{f'(x_n)}$$

may be written

$$x_{n+1} = \phi(x_n).$$

We observed (see p. 38) that in general the iteration method converges when $|\phi'(x)| < 1$ near the root. In the Newton-Raphson case.

$$\phi'(x) = 1 - \frac{[f'(x)]^2 - f(x)f''(x)}{[f'(x)]^2} = \frac{f(x)f''(x)}{[f'(x)]^2},$$

so that the criterion for convergence is

$$|f(x)f''(x)| < [f'(x)]^2 \ ;$$

convergence is *not* so assured as for the bisection method (say).

4 SPEED OF CONVERGENCE

Since

$$x_{n+1} = x_n - \frac{f(x_n)}{f'(x_n)},$$

we have

$$e_{n+1} = \alpha - x_{n+1} = \alpha - x_n + \frac{f(x_n)}{f'(x_n)}$$

$$= e_n + \frac{f(\alpha - e_n)}{f'(\alpha - e_n)}$$

$$= e_n + \frac{f(\alpha) - e_n f'(\alpha) + (e_n^2/2)f''(\alpha) - \ldots}{f'(\alpha) - e_n f''(\alpha) + (e_n^2/2)f'''(\alpha) - \ldots}$$

$$= e_n - e_n + \frac{e_n^2 \, f''(\alpha)}{2 \, f'(\alpha)} + \ldots$$

$$\approx + \frac{e_n^2 \, f''(\alpha)}{2 \, f'(\alpha)}, \text{ when } e_n \text{ is sufficiently small.}$$

This result states that the error at step $n + 1$ is proportional to the *square* of the error at step n; hence (if $f''(\alpha) \approx 4f'(\alpha)$) an answer correct to 1 decimal place at one step should be accurate to 2 places at the next step, 4 at the next, 8 at the next, etc. This quadratic ('second order') convergence outstrips the rate of convergence of the methods of bisection and false position.

In relatively little used computer programs, it may be wise to prefer the methods of bisection or false position, since convergence is virtually assured. For desk work or for computer routines in constant use, the Newton-Raphson method is usually preferred.

5 THE SQUARE ROOT

One application of the Newton-Raphson method is in the computation of square roots. Now, finding $\sqrt{a}$ is equivalent to finding the positive root of $x^2 = a$ or

$$f(x) = x^2 - a = 0.$$

Since $f'(x) = 2x$. we have the Newton-Raphson iteration formula:

$$x_{n+1} = x_n - \frac{x_n^2 - a}{2x_n}$$

$$= \tfrac{1}{2}(x_n + \frac{a}{x_n}).$$

(As mentioned in the Prologue, this was known to the ancient Greeks.) Thus, if $a = 16$ and $x_0 = 5$, we have $x_1 = \tfrac{1}{2}(5 + 3\cdot2) = 4\cdot1$, $x_2 = \tfrac{1}{2}(4\cdot1 + 3\cdot902) = 4\cdot001$ and $x_3 = \tfrac{1}{2}(4\cdot001 + 3\cdot999) = 4\cdot000$, working to 3D.

Checkpoint

---

1. What is the geometrical interpretation of the Newton-Raphson iterative procedure?

2. What is the convergence criterion for the Newton-Raphson method?

3. What major advantage has the Newton-Raphson method over some other methods?

EXERCISES

1. Use the Newton-Raphson method to solve for the (positive) root of $3xe^x = 1$ to four significant figures.
   [*Note*: Tables of natural (Napierian) logarithms are more readily available than tables of the exponential, so that one might prefer to solve the equivalent equation $f(x) = \log_e 3x + x = \log_e 3 + \log_e x + x = 0.$]

2. Derive the Newton-Raphson iteration formula

$$x_{n+1} = x_n - \frac{x_n^k - a}{kx_n^{k-1}}$$

for finding the kth root of a.

3. Compute the square root of 10 to 5 significant figures, from the initial guess 1.

4. Use the Newton-Raphson method to find (to 4*D*) the root of the equation

$$x + \cos x = 0.$$

SYSTEMS OF LINEAR EQUATIONS 1
Solution by elimination

Many physical systems can be modelled by a set of linear equations which describe relationships between system variables. In simple cases there are two or three variables; in complex systems (for example, in a linear model of the economy of a country) there may be several hundred variables. Linear systems also arise in connection with many problems of numerical analysis. Examples of these are the solution of differential equations by finite difference methods, statistical regression analysis and the solution of eigenvalue problems (see for example Conte and de Boor).

It is necessary, therefore, to have available rapid and accurate methods for solving systems of linear equations. The student will already be familiar with solving systems of equations with 2 or 3 variables by elimination methods. In this Step we shall give a formal description of the *Gauss elimination method* for *n*-variable systems and discuss certain errors which might arise in solutions. An addition to the *Gauss* method which helps to reduce loss of accuracy will be described in the next Step.

1 NOTATION AND DEFINITIONS

i) *An example in 3 variables*

$$x + y - z = 2$$
$$x + 2y + z = 6$$
$$2x - y + z = 1$$

This is a set of 3 linear equations in the 3 *variables* (or *unknowns*) x, y and z. By 'solution of the system' we mean the determination of a set of values for x, y, and z which satisfies each one of the equations. In other words, if values (X, Y, Z) satisfy all equations simultaneously, then (X, Y, Z) constitute a solution of the system.

ii) *The general system of n equations, n variables*

The general system in n variables may be written as follows:

$$\left.\begin{array}{l} a_{11}x_1 + a_{12}x_2 + \ldots + a_{1n}x_n = b_1 \\ a_{21}x_1 + a_{22}x_2 + \ldots + a_{2n}x_n = b_2 \\ \qquad . \quad . \quad . \quad . \quad . \quad . \quad . \quad . \quad . \quad . \\ a_{n1}x_1 + a_{n2}x_2 + \ldots + a_{nn}x_n = b_n \end{array}\right\} n \text{ equations}$$

The dots indicate, of course, similar terms in the variables x_3, x_4 etc., and the remaining $(n-3)$ equations which complete the system.

In this notation, the *variables* are denoted by x_1, x_2, ..., x_n; sometimes we write x_i, $i = 1, \ldots, n$ to represent the variables. The *coefficients* of the variables may be detached and written in a *matrix* thus:

$$\mathbf{A} = \begin{bmatrix} a_{11} & a_{12} & \cdots & a_{1n} \\ a_{21} & a_{22} & \cdots & a_{2n} \\ . & . & . & . \\ a_{n1} & a_{n2} & \cdots & a_{nn} \end{bmatrix}$$

The notation a_{ij} will be used to denote the coefficient of x_j in the ith equation. Note that it occurs in the *ith row* and *jth column* of the matrix.

The numbers on the right-hand side of the equations are called *constants*, and may be written in a column vector, thus:

$$\mathbf{b} = \begin{bmatrix} b_1 \\ b_2 \\ . \\ . \\ . \\ b_n \end{bmatrix}$$

The coefficient matrix may be combined with the constant vector to form the *augmented matrix*, thus:

$$\begin{bmatrix} a_{11} & a_{12} & \cdots & a_{1n} & b_1 \\ a_{21} & a_{22} & \cdots & a_{2n} & b_2 \\ . & . & . & . & . \\ a_{n1} & a_{n2} & \cdots & a_{nn} & b_n \end{bmatrix}$$

It is usual to work directly with the augmented matrix when using elimination methods of solution.

2 THE EXISTENCE OF SOLUTIONS

For any particular system of n linear equations there may be a single solution $(X_1, X_2, \dots X_n)$, or no solution, or an infinity of solutions. In the theory of linear algebra, theorems are given and conditions stated which enable us to decide which category a given system falls into. We shall not treat the question of existence of solutions in this book, but for the benefit of readers familiar with matrices and determinants we will state the following theorem.

THEOREM: A linear system of n equations in n variables, with coefficient matrix $\mathbf{A}$ and constants vector $\mathbf{b} \neq \mathbf{0}$, has a unique solution if and only if the determinant of $\mathbf{A}$ is not zero.

If all elements of $\mathbf{b}$ are zero, then the system has solution $\mathbf{x} = \mathbf{0}$. It has no other solution unless the determinant of $\mathbf{A}$ is zero, in which case it has an infinite number of solutions.

3 GAUSS ELIMINATION METHOD

Gauss method of elimination is to transform the given system of equations into an equivalent system which is in *triangular form*; this new form can be solved easily by a process called *back-substitution*. We shall demonstrate the process by solving the example of Section 1(i).

a) *Transformation to triangular form*

$$\begin{aligned} x + \ y - \ z &= 2 \qquad (1) \\ x + 2y + \ z &= 6 \qquad (2) \\ 2x - \ y + \ z &= 1 \qquad (3) \end{aligned}$$

First stage: eliminate x from equations (2) and (3), using equation (1).

$$\begin{aligned} x + \ y - \ z &= 2 \qquad &(1)'. \\ y + 2z &= 4 \qquad &(2)' \text{ (equation (2) minus equation (1))} \\ - 3y + 3z &= -3 \qquad &(3)' \text{ (equation (3) minus twice equation (1))} \end{aligned}$$

Second stage: eliminate y from (3)', using (2)'.

$$\begin{aligned} x + \ y - \ z &= 2 \qquad &(1)'' \\ y + 2z &= 4 \qquad &(2)'' \\ 9z &= 9 \qquad &(3)'' \text{ (equation (3)' plus three times (2)')} \end{aligned}$$

The system is now in *triangular form*. The coefficient matrix is

$$\begin{bmatrix} 1 & 1 & -1 \\ 0 & 1 & 2 \\ 0 & 0 & 9 \end{bmatrix} \text{, an upper triangular matrix.}$$

b) *Solution by back-substitution*

The system in triangular form is easily solved by obtaining z from $(3)''$, then y from $(2)''$ and finally x from $(1)''$. This procedure is called back-substitution.

Thus, $\quad z = 1 \qquad\qquad$ dividing $(3)''$ by 9 ;

$\qquad\qquad y = 4 - 2z \qquad$ from $(2)''$

$\qquad\qquad\quad = 2 \qquad\qquad$ using $z = 1$;

$\qquad\qquad x = 2 - y + z \qquad$ from $(1)''$

$\qquad\qquad\quad = 1 \qquad\qquad$ using $z = 1$ and $y = 2$

4   THE TRANSFORMATION OPERATIONS

When transforming a system to triangular form we use one or more of the following *elementary operations* at every step:

1) multiplication of an equation by a constant;
2) addition to one equation of some multiple of another equation;
3) interchange of two equations.

Mathematically speaking, it should be clear to the reader that performing elementary operations on a system of linear equations leads to equivalent systems which have the same solutions. This statement requires proof; this may be found as a theorem in books on linear algebra (see bibliography). It forms the basis of all elimination methods for solving systems of linear equations.

5 GENERAL TREATMENT OF THE ELIMINATION PROCESS

We shall now describe the elimination process as applied to three equations written in general notation. We shall begin with the augmented matrix, and show the multipliers necessary (in the column headed m) to perform the transforming operations.

multipliers
m   *Augmented matrix*

$$\begin{array}{l} \\ m_1 = -a_{21}/a_{11} \\ m_2 = -a_{31}/a_{11} \end{array} \begin{bmatrix} a_{11} & a_{12} & a_{13} & b_1 \\ a_{21} & a_{22} & a_{23} & b_2 \\ a_{31} & a_{32} & a_{33} & b_3 \end{bmatrix} \begin{array}{l} (1) \\ (2) \\ (3) \end{array}$$

First stage: eliminate the coefficients a_{21} and a_{31}, using row (1)

$$m \quad \begin{bmatrix} a_{11} & a_{12} & a_{13} & b_1 \\ 0 & a'_{22} & a'_{23} & b'_2 \\ 0 & a'_{32} & a'_{33} & b'_3 \end{bmatrix} \begin{matrix} (1)' \\ (2)' \ (\text{row } (2) + m_1 \text{ times row } (1)) \\ (3)' \ (\text{row } (3) + m_2 \text{ times row } (1)) \end{matrix}$$

$m_3 = -a'_{32}/a'_{22}$

Second stage: eliminate a'_{32}, using row (2)'.

$$\begin{bmatrix} a_{11} & a_{12} & a_{13} & b_1 \\ 0 & a'_{22} & a'_{23} & b'_2 \\ 0 & 0 & a''_{33} & b''_3 \end{bmatrix} \begin{matrix} (1)'' \\ (2)'' \\ (3)'' \ (\text{row } (3)' + m_3 \text{ times row } (2)') \end{matrix}$$

The matrix is now in the form necessary for back-substitution to be carried out. The full system of equations at this point is

$$a_{11}x_1 + a_{12}x_2 + a_{13}x_3 = b_1$$
$$a'_{22}x_2 + a'_{23}x_3 = b'_2,$$
$$a''_{33}x_3 = b''_3.$$

It is equivalent to the original system. The solution procedure continues as follows:

$$x_3 = b''_3/a''_{33},$$
$$x_2 = (b'_2 - a'_{23}x_3)/a'_{22},$$
$$x_1 = (b_1 - a_{12}x_2 - a_{13}x_3)/a_{11}.$$

Notes

i)   The diagonal elements $a_{11}, a'_{22}, a''_{33}$ used in the successive eliminations are called *pivot elements*.

ii)   To proceed from one stage to the next it is necessary for the pivot element to be non-zero (notice that the pivot elements are used as divisors in the multipliers, and in the final solution). If at any stage a pivot element vanishes we rearrange the remaining rows of the matrix so as to obtain a non-zero pivot; if this is not possible, then the system of linear equations has no solution.

iii)   If a pivot element is small compared with the elements in its column which have to be eliminated, the multipliers used at that stage will be greater than one in magnitude. The use of large multipliers in the elimination and back-substitution processes leads to a magnification of round-off errors. A method of avoiding small pivot elements is given in the next Step.

iv) When carrying out an elimination method, an extra check column, containing row-sums, should be computed at each stage (see the example below). Its elements are treated in exactly the same way as the equation coefficients. After each stage is completed the new row sums should equal the new check column elements (within roundoff error).

6 NUMERICAL EXAMPLE, WITH CHECK COLUMN

System

$$0.34x_1 - 0.58x_2 + 0.94x_3 = 2.0$$

$$0.27x_1 + 0.42x_2 + 0.13x_3 = 1.5$$

$$0.20x_1 - 0.51x_2 + 0.54x_3 = 0.8$$

Solution

The working is set out in tabular form below. The multipliers were rounded to $3S$ before use; the final results from all other calculations were rounded to $2S$. Working with so few significant figures leads to considerable errors in the solution, as is shown below by an examination of *residuals*.

	m	Augmented matrix				Check Column	
		0.34	−0.58	0.94	2.0	2.70	
		0.27	0.42	0.13	1.5	2.32	
		0.20	−0.51	0.54	0.8	1.03	
First stage		0.34	−0.58	0.94	2.0		
	−0.794		0.88	−0.62	−0.088	0.18	(0.17)
	−0.588		−0.17	−0.01	−0.38	−0.56	(−0.56)
Second stage		0.34	−0.58	0.94	2.0		
			0.88	−0.62	−0.088		
	0.193			−0.13	−0.40	−0.53	(−0.53)

Back substitution

$$-0.13x_3 = -0.40 \rightarrow x_3 \approx 3.1$$

$$0.88x_2 - 0.62 \times 3.1 = -0.088 \rightarrow x_2 \approx 2.1$$

$$0.34x_1 - 0.58 \times 2.1 + 0.94 \times 3.1 = 2.0 \rightarrow x_1 \approx 0.89$$

Check

Sum the original three equations: $0.81x_1 - 0.67x_2 + 1.61x_3 = 4.3$.
Insert the solution: $0.81 \times 0.89 - 0.67 \times 2.1 + 1.61 \times 3.1 = 4.3049$.

Residuals

In order to judge the accuracy of the solution, we may insert the solution into the left hand side of each of the original equations, and compare the results with the right-hand side constants. The differences between the results and the constants are called *residuals*. For the example:

$$\text{residuals}$$
$$0.34 \times 0.89 - 0.58 \times 2.1 + 0.94 \times 3.1 = 1.9986\,; 2.00 - 1.9986 = \quad 0.0014,$$
$$0.27 \times 0.89 + 0.42 \times 2.1 + 0.13 \times 3.1 = 1.5253\,; 1.50 - 1.5253 = -0.0253,$$
$$0.20 \times 0.89 - 0.51 \times 2.1 + 0.54 \times 3.1 = 0.781\;\;; 0.80 - 0.781 = \quad 0.019.$$

It would seem reasonable to believe that if the residuals are small the solution is a good one. This is usually the case. Sometimes, however, small residuals are *not* indicative of a good solution. This point is taken up under 'ill-conditioning', in the next Step.

Checkpoint

---

1. When transforming the augmented matrix, what kinds of operation are permissible?
2. What is the final form of the coefficient matrix, before back-substitution begins?
3. What are pivot elements? Why must small pivot elements be avoided if possible?

EXERCISES

Solve the following systems by Gauss elimination. Use a check column. Compute the residuals.

1. $x_1 + x_2 - x_3 = 0$

     $2x_1 - x_2 + x_3 = 6$

     $3x_1 + 2x_2 - 4x_3 = -4$

2. $5 \cdot 6x + 3 \cdot 8y + 1 \cdot 2z = 1 \cdot 4$

     $3 \cdot 1x + 7 \cdot 1y - 4 \cdot 7z = 5 \cdot 1$

     $1 \cdot 4x - 3 \cdot 4y + 8 \cdot 3z = 2 \cdot 4$

3.   (i) $2x + 6y + 4z = 5$

         $6x + 19y + 12z = 6$

         $2x + 8y + 14z = 7$

   (ii) $1 \cdot 3x + 4 \cdot 6y + 3 \cdot 1z = -1$

         $5 \cdot 6x + 5 \cdot 8y + 7 \cdot 9z = 2$

         $4 \cdot 2x + 3 \cdot 2y + 4 \cdot 5z = -3$

SYSTEMS OF LINEAR EQUATIONS 2
Errors and ill-conditioning

For any system of linear equations, the question of how much error there may be in a solution obtained by a numerical method is a very difficult one to answer. A general discussion of the problems it raises is beyond the scope of this book. However, some of the sources of errors in solutions will be indicated.

1 UNCERTAINTY IN THE COEFFICIENTS AND CONSTANTS

In many practical cases the coefficients of the variables, and also the constants on the right-hand sides of the equations, are obtained from observations of experiments or from other numerical calculations. They are not known with certainty; and therefore when the solution of the system is found, it too will contain a measure of uncertainty. To show how this kind of error is carried through in calculations, we shall solve a simple example in two variables, assuming that the constants are both uncertain to $\pm\ 0{\cdot}01$.

$$2x + y = 4\,(\pm\ 0{\cdot}01)$$
$$-\,x + y = 1\,(\pm\ 0{\cdot}01)$$

Solving by Gauss elimination and back-substitution:

$$2x + y = 4\,(\pm\ 0{\cdot}01),$$
$$\tfrac{3}{2}y = 1\,(\pm\ 0{\cdot}01) + 2\,(\pm\ 0{\cdot}005).$$

Therefore $\qquad \tfrac{3}{2}y$ lies between $2{\cdot}985$ and $3{\cdot}015,$

so $\qquad\qquad y$ lies between $1{\cdot}990$ and $2{\cdot}010.$

From the first equation we now obtain

$$2x = 4\,(\pm\ 0{\cdot}01) - 2\,(\pm\ 0{\cdot}01),$$

so x lies between 0·99 and 1·01.

If the system were exact in its coefficients and constants, its exact solution would be $x = 1$, $y = 2$. Since the constants are not known exactly, it is meaningless to talk of an exact solution; the best that can be said is that $0.99 \leqslant x \leqslant 1.01$ and $1.99 \leqslant y \leqslant 2.01$.

In this example the uncertainty in the solution is of the same order as that in the constants. Generally, however, the uncertainty in the solutions is greater than that in the constants.

2 ROUND-OFF ERRORS

Any numerical method for solving systems of linear equations involves large numbers of arithmetic operations. For example, in the Gauss elimination method of the previous step, it may be shown (see for example Conte and de Boor) that there are $n(n + 1)/2$ divisions, $\left(\dfrac{n^3}{3} + \dfrac{n^2}{2} - \dfrac{5n}{6}\right)$ multiplications and $\left(\dfrac{n^3}{3} + \dfrac{n^2}{2} - \dfrac{5n}{6}\right)$ additions required to arrive at the solution of a system which has n unknowns.

Since round-off errors are propagated at each step of an algorithm, the growth of round-off errors can be such as to lead to a solution very far from the true one when n is large.

3 PIVOTAL CONDENSATION

In the Gauss elimination method, the build-up of round-off errors may be reduced by so arranging the equations that the use of large multipliers in the elimination operations is avoided. The procedure to be carried out is known as *pivotal condensation* (or Gauss elimination with partial pivoting). The general rule to follow is: at each elimination stage, arrange the rows of the augmented matrix so that the new pivot element is larger in absolute value than (or equal to) any element beneath it in its column.

Use of this rule ensures that the multipliers used at each stage have magnitude less than or equal to one. To show the rule in operation we treat a simple example, using exact arithmetic.

Solve

$$2x + 5y + 8z = 36$$
$$4x + 8y - 12z = -16$$
$$x + 8y + z = 20$$

Tabular solution is as follows, the pivot elements being printed in bold-face numerals.

Stage	Multipliers	Augmented matrix	Explanation
1. Eliminate x–terms in two equations; note that both multipliers have magnitude less than 1.	$-1/2$ $-1/4$	**4** 8 -12 -16 2 5 8 36 1 8 1 20	The 1st and 2nd equations have been interchanged; the pivot element 4 is now the largest in the x–column.
2. Eliminate the y–term in the third equation.		**4** 8 -12 -16 0 1 14 44 0 **6** 4 24	Rows 2 and 3 must be interchanged, so that 6 becomes the next pivot element rather than 1.
Note that $\|m\| < 1$	$-1/6$	**4** 8 -12 -16 0 **6** 4 24 0 0 40/3 40	
3. Solve by back-substitution: $z = 3$, $y = 2$, $x = 1$.			

4 ILL-CONDITIONING

Certain systems of linear equations are such that their solutions are very sensitive to small changes (and therefore to errors) in their coefficients and constants. We give an example below in which 1% changes in two coefficients change the solution by a factor of 10 or more. Such systems are said to be *ill-conditioned*. If a system is ill-conditioned, a solution obtained by a numerical method may be very different from the exact solution, even though great care is taken to keep round-off and other errors very small.

Example

$$2x + \quad y = 4$$

$$2x + 1{\cdot}01y = 4{\cdot}02$$

This has the exact solution $x = 1$, $y = 2$.

Making 1% changes in the coefficients of the second equation and a 5% change in the constant of the first, gives the system

$$2x + y = 3\cdot82,$$
$$2\cdot02x + y = 4\cdot02.$$

It is easily verified that the exact solution to *this* system is $x = 10$, $y = -16\cdot18$. This is very different from the solution to the first system. Both these systems are therefore ill-conditioned.

If a system is ill-conditioned, the usual procedure of checking a numerical solution – by inserting it in the original equations (or a combination of them) and 'seeing how well it fits' – is not valid. There are more precise ways of checking solutions, and many tests have been proposed for determining whether or not a system is ill-conditioned; for these the reader is referred to more advanced texts on Numerical Analysis.

Checkpoint

1. Describe the types of error that may affect the solution of a system of linear equations.
2. How can use of the pivotal condensation rule contribute to a reduction of errors?
3. Is it true to say that an ill-conditioned system has not got an exact solution?

EXERCISES

1. Find the range of solutions for the following system, assuming uncertainties in the constants as shown:
$$x - y = 1\cdot4\,(\pm\,0\cdot01),$$
$$x + y = 3\cdot8\,(\pm\,0\cdot05).$$

2. Solve the following systems by Gauss elimination:

(i) $\quad x - 10y = -21\cdot8,$
$\quad 10x + y = 14\cdot3\,;$

(ii) $\quad x + 5y - z = 4,$
$\quad 2x - y + 3z = 7,$
$\quad 3x - y + 5z = 12\,;$

(iii) $2\cdot1x_1 + 2\cdot4x_2 + 8\cdot1x_3 = 62\cdot76,$
$\quad 7\cdot2x_1 + 8\cdot5x_2 - 6\cdot3x_3 = -1\cdot93,$
$\quad 3\cdot4x_1 - 6\cdot4x_2 + 5\cdot4x_3 = 16\cdot24.$

3. Using floating-point arithmetic with a four-decimal-place mantissa, solve the following system with and without using pivotal condensation. Compare your answers with the exact answer, which is $x = 1 \cdot 000$, $y = 0 \cdot 5000$.

$$0 \cdot 2310 \times 10^{-2}x + 0 \cdot 4104 \times 10^{-1}y = 0 \cdot 2283 \times 10^{-1}$$
$$0 \cdot 4200 \times 10^{0}x + 0 \cdot 5368 \times 10^{1}y = 0 \cdot 3104 \times 10^{1}$$

4. Show that for a linear system of 3 unknowns, the Gauss elimination procedure requires 3 divisions, 8 multiplications and 8 additions to complete the triangularization; and a further 3 divisions, 3 multiplications, and 3 additions to carry out the back-substitution.

    Derive the general formulae given in Section 2 for numbers of required arithmetic operations.

5. Study the ill-conditioning example given in Section 4 in the following ways.
    i)   Plot the lines of the first system on graph paper; now describe ill-conditioning in geometrical terms when only two unknowns are involved.
    ii) Insert the solution of the first system into the left-hand side of the second system. Does $x = 1$, $y = 2$ 'look like' a good solution to the second system? Comment.
    iii) Insert the solution of the second system into the left-hand side of the first system. Comment.

6. (This is an example of ill-conditioning due to T.S. Wilson.)

$$10x_1 + 7x_2 + 8x_3 + 7x_4 = 32$$
$$7x_1 + 5x_2 + 6x_3 + 5x_4 = 23$$
$$8x_1 + 6x_2 + 10x_3 + 9x_4 = 33$$
$$7x_1 + 5x_2 + 9x_3 + 10x_4 = 31$$

Insert the 'solution' $(6 \cdot 0, -7 \cdot 2, 2 \cdot 9, -0 \cdot 1)$ into the left-hand side. Would you claim this solution to be a good one? Now insert the solution $(1 \cdot 0, 1 \cdot 0, 1 \cdot 0, 1 \cdot 0)$. Comment on the dangers of making claims!

SYSTEMS OF LINEAR EQUATIONS 3
The Gauss-Seidel iterative method

The methods used in the previous Steps for solving systems of linear equations are termed direct methods. When a *direct method* is used, and if round-off and other errors do not arise, an exact solution is reached after a finite number of arithmetic operations. In general, of course, round-off errors do arise; and when large systems are being solved by direct methods the growth of errors can be such as to lead to useless results.

1 ITERATIVE METHODS

Iterative methods provide an alternative approach. Recall that an *iterative method* starts with an approximate solution, and uses it in a recurrence formula to provide another approximate solution; by repeatedly applying the formula, a sequence of solutions is obtained which (under suitable conditions) converges to the exact solution. Iterative methods have the advantages of simplicity of operation and ease of implementation on computers, and they are relatively insensitive to propagation of errors; they would be used in preference to direct methods for solving linear systems involving several hundred variables, particularly if many of the coefficients were zero. Systems of up to 10 000 variables have been successfully solved on computers by iterative methods, whereas systems of 500 or more variables are difficult or impossible to solve by direct methods.

2 THE GAUSS-SEIDEL METHOD[†]

Only one iterative method for linear equations, due to Gauss and improved by Seidel, will be presented in this text. Consider the following.

† This algorithm is suitable for automatic computation. A flow-chart for study and use in programming may be found on page 153.

Example

Solve the system

$$10x_1 + 2x_2 + x_3 = 13,$$
$$2x_1 + 10x_2 + x_3 = 13,$$
$$2x_1 + x_2 + 10x_3 = 13.$$

We shall use the *iterative method.*

The first step is to solve the first equation for x_1, the second for x_2 and the third for x_3. This transforms the system to the following:

$$x_1 = 1\cdot3 - 0\cdot2x_2 - 0\cdot1x_3 \quad (1)$$
$$x_2 = 1\cdot3 - 0\cdot2x_1 - 0\cdot1x_3 \quad (2)$$
$$x_3 = 1\cdot3 - 0\cdot2x_1 - 0\cdot1x_2 \quad (3)$$

An initial solution is now assumed; we shall use $x_1 = 0$, $x_2 = 0$, $x_3 = 0$. Inserting this into the right-hand side of (1) gives $x_1 = 1\cdot3$. This value for x_1 is used immediately together with the remainder of the initial solution (i.e. $x_2 = 0$ and $x_3 = 0$) in the right-hand side of (2), giving $x_2 = 1\cdot3 - 0\cdot2 \times 1\cdot3 - 0 = 1\cdot04$. Finally, $x_1 = 1\cdot3$ and $x_2 = 1.04$ are inserted in (3) to produce $x_3 = 0\cdot936$. This completes the first iteration; we have obtained a second approximate solution $(1\cdot3, 1\cdot04, 0\cdot936)$.

Beginning with the second solution, we can repeat the process to obtain a third. Clearly we can continue in this way, and obtain a sequence of approximate solutions. Under certain conditions on the coefficients of the system, the sequence will converge to the exact solution.

We can set up recurrence relations which show clearly how the *iterative process* proceeds. Using $(x_1^{(k+1)}, x_2^{(k+1)}, x_3^{(k+1)}$ and $(x_1^{(k)}, x_2^{(k)}, x_3^{(k)})$ to denote the $(k+1)$th and kth solutions respectively, we have

$$x_1^{(k+1)} = 1\cdot3 - 0\cdot2x_2^{(k)} - 0\cdot1x_3^{(k)} \quad (1)'$$
$$x_2^{(k+1)} = 1\cdot3 - 0\cdot2x_1^{(k+1)} - 0\cdot1x_3^{(k)} \quad (2)'$$
$$x_3^{(k+1)} = 1\cdot3 - 0\cdot2x_1^{(k+1)} - 0\cdot1x_2^{(k+1)} \quad (3)'$$

We begin with $(x_1^{(0)}, x_2^{(0)}, x_3^{(0)}) = (0, 0, 0)$ and then apply these relations repeatedly in the order $(1)', (2)', (3)'$. Note that when we insert values for x_1, x_2 and x_3 into the right-hand sides we always use the most recent estimates found for each unknown

3 CONVERGENCE

The sequence of solutions produced by the iterative process may be displayed in a table, thus:

Iteration k	Approximate solution (Gauss-Seidel)		
	$x_1^{(k)}$	$x_2^{(k)}$	$x_3^{(k)}$
0	0	0	0
1	1·3	1·04	0·936
2	0·9984	1·00672	0·999648
3	0·998691	1·000297	1·000232

The reader may check that the exact solution for this system is (1, 1, 1). It is seen that the Gauss-Seidel solutions are rapidly approaching this; in other words, the method is converging.

In practice, of course, the exact solution is not known. It is customary to end the iterative procedure as soon as the differences between the $x^{(k+1)}$ values and the $x^{(k)}$ values are suitably small. One stopping-rule is to end the iteration when first $S_k = \Sigma_i |x_i^{(k+1)} - x_i^{(k)}|$ is less than a prescribed small number (e.g. 0·00001).

The question of convergence with a given system of equations is a crucial one; as in the above example, the Gauss-Seidel method may quickly lead to a solution very close to the exact one; on the other hand it may converge too slowly to be of practical use, or it may produce a sequence which diverges from the exact solution. The reader is referred to more advanced texts (see for example Conte and de Boor) for treatments of this question.

To improve the chance (and rate) of convergence, before applying the iterative method the system of equations should be arranged so that as far as possible each leading-diagonal coefficient is the largest (in absolute value) in its row.

Checkpoint

1. What is an essential difference between a direct method and an iterative method?
2. Give some advantages of the use of iterative methods rather than direct methods.

3. How can the chance of success with the Gauss-Seidel method be improved?

EXERCISES

1. For the example treated above, compute the value of S_3, the quantity used in the suggested stopping rule after the third iteration.
2. Solve the following systems by the Gauss-Seidel method.

   i) (to 4D)

   $$x - y + 10z = -7$$
   $$20x + 3y - 2z = 51$$
   $$2x + 8y + 4z = 25 \qquad \text{(remember to rearrange)}$$

   Compute the value of S_k (to 5D) after each iteration.

   i (to 4D)

   $$10x - y = 1$$
   $$-x + 10y - z = 1$$
   $$- y + 10z - w = 1$$
   $$- z + 10w = 1$$

SYSTEMS OF LINEAR EQUATIONS 4*
Matrix inversion

In this optional Step it is assumed that the reader is familiar with elementary matrix algebra. The general system of n linear equations in n variables (see Step 11, Section 1) can be written in matrix form $Ax = b$, and we seek a vector x which satisfies this equation.

1 THE INVERSE MATRIX

Provided that the determinant of A is non-zero, there exists a matrix called the *inverse of* A (denoted by A^{-1}) which is such that the matrix product of A^{-1} and A is equal to the $n \times n$ unit matrix I. That is, $A^{-1}A = I = AA^{-1}$.

Multiplying the equation $Ax = b$ from the left by the inverse matrix A^{-1} we obtain $A^{-1}Ax = A^{-1}b$; and this gives $x = A^{-1}b$ (since $A^{-1}A = I$, and $Ix = x$).

Thus the solution to the system of linear equations can be obtained by first finding the inverse of the coefficient matrix A, and then forming the product $A^{-1}b$.

However, this method of solution is not to be recommended generally. The problem of finding the inverse matrix is itself a numerical one, which requires for its solution many more operations (and therefore involves more round-off and other errors) than any of the methods described in previous Steps.

It would be sensible to compute the inverse first if either (a) the inverse were required for some additional reason (e.g. it may contain special statistical information or be of use in some other formula or calculation), or (b) there were several linear systems to be solved, each having the same coefficient matrix.

2 METHOD FOR INVERTING A MATRIX

There are many numerical methods for finding the inverse of a matrix. We shall describe one which uses the Gauss elimination and back-substitution procedures of Step 11. It is simple to apply and is computationally efficient. We shall illustrate the method by applying it to a 2×2 matrix and a 3×3 matrix; it should then be clear to the reader how the method may be extended for use with $n \times n$ matrices.

i) 2×2 *example*

Let

$$\mathbf{A} = \begin{bmatrix} 2 & 1 \\ 4 & 5 \end{bmatrix}.$$

It is desired to find a matrix $\mathbf{A}^{-1} = \begin{bmatrix} u_1 & u_2 \\ v_1 & v_2 \end{bmatrix}$ such that

$$\mathbf{A}\mathbf{A}^{-1} = \mathbf{I} = \begin{bmatrix} 1 & 0 \\ 0 & 1 \end{bmatrix}$$

This is equivalent to solving the two systems $\mathbf{A} \begin{bmatrix} u_1 \\ v_1 \end{bmatrix} = \begin{bmatrix} 1 \\ 0 \end{bmatrix}$ and $\mathbf{A} \begin{bmatrix} u_2 \\ v_2 \end{bmatrix} = \begin{bmatrix} 0 \\ 1 \end{bmatrix}$

   The method proceeds as follows:
a)   Form the augmented matrix

$$[\mathbf{A}|\mathbf{I}] = \begin{bmatrix} 2 & 1 & 1 & 0 \\ 4 & 5 & 0 & 1 \end{bmatrix}$$

b)   Apply elementary row operations to the augmented matrix such that $\mathbf{A}$ is transformed to an upper triangular matrix $\tilde{\mathbf{A}}$ (see Step 11, Section 5):

$$\begin{array}{cc} \mathbf{A} & \mathbf{I} \end{array} \qquad\qquad \begin{array}{cc} \tilde{\mathbf{A}} & \tilde{\mathbf{I}} \end{array}$$

$$\begin{bmatrix} 2 & 1 & 1 & 0 \\ 4 & 5 & 0 & 1 \end{bmatrix} \rightarrow \begin{bmatrix} 2 & 1 & 1 & 0 \\ 0 & 3 & -2 & 1 \end{bmatrix}$$
$$\text{(row 2 — twice row 1)}$$

c)   Solve the two systems

$$\begin{bmatrix} 2 & 1 \\ 0 & 3 \end{bmatrix}\begin{bmatrix} u_1 \\ v_1 \end{bmatrix} = \begin{bmatrix} 1 \\ -2 \end{bmatrix} \quad \text{and} \quad \begin{bmatrix} 2 & 1 \\ 0 & 3 \end{bmatrix}\begin{bmatrix} u_2 \\ v_2 \end{bmatrix} = \begin{bmatrix} 0 \\ 1 \end{bmatrix},$$

using the back-substitution method. Note how the systems are constructed, using $\tilde{A}$ and columns of $\tilde{I}$. Then, from the first system, $3v_1 = -2$, $v_1 = -\frac{2}{3}$, and $2u_1 + v_1 = 1$, so $2u_1 = 1 + \frac{2}{3}$, $u_1 = \frac{5}{6}$.

From the second system, $3v_2 = 1$, $v_2 = \frac{1}{3}$, and $2u_2 + v_2 = 0$, so $2u_2 = -\frac{1}{3}$, $u_2 = -\frac{1}{6}$. The required inverse matrix is

$$\mathbf{A}^{-1} = \begin{bmatrix} u_1 & u_2 \\ v_1 & v_2 \end{bmatrix} = \begin{bmatrix} \frac{5}{6} & -\frac{1}{6} \\ -\frac{2}{3} & \frac{1}{3} \end{bmatrix}$$

d) Check: $\mathbf{AA}^{-1}$ should equal $\mathbf{I}$. By multiplication we find

$$\begin{bmatrix} 2 & 1 \\ 4 & 5 \end{bmatrix} \begin{bmatrix} \frac{5}{6} & -\frac{1}{6} \\ -\frac{2}{3} & \frac{1}{3} \end{bmatrix} = \begin{bmatrix} 1 & 0 \\ 0 & 1 \end{bmatrix},$$

so $\mathbf{A}^{-1}$ is correct.

In this simple example it has been possible to work with fractions, so no round-off errors occur and the resulting inverse matrix is exact.

ii) *A 3 × 3 example*

A check column containing row sums should be maintained whilst the transformation of the augmented matrix is being carried out. To avoid excessive loss of significance due to accumulation of round-off errors, we may retain an extra one or two significant figures in all calculations. The final result should be checked by computing $\mathbf{AA}^{-1}$, which should be approximately equal to the unit matrix I.

As an example we shall find the inverse matrix $\mathbf{A}^{-1}$ of

$$\mathbf{A} = \begin{bmatrix} 0{\cdot}20 & 0{\cdot}24 & 0{\cdot}12 \\ 0{\cdot}10 & 0{\cdot}24 & 0{\cdot}24 \\ 0{\cdot}05 & 0{\cdot}30 & 0{\cdot}49 \end{bmatrix}$$

To show the effects of errors we shall assume that the elements of A are correct to $2S$ and in the calculation of $\mathbf{A}^{-1}$ we shall work to $2S$. The results of the calculations may be conveniently displayed in tabular form, as follows.

Multipliers	$\mathbf{A}$ transforms to $\widetilde{\mathbf{A}}$			$\mathbf{I}$ transforms to $\widetilde{\mathbf{I}}$			Check Σ	Row operations
	0·20	0·24	0·12	1	0	0	1·56	(1)
·	0·10	0·24	0·24	0	1	0	1·58	(2)
	0·05	0·30	0·49	0	0	1	1·84	(3)
	0·20	0·24	0·12	1	0	0		(4)=(1)
−0·5	0	0·12	0·18	−0·5	1	0	0·80	(5)=(2)−0·5(1)
−0·25	0	0·24	0·46	−0·25	0	1	1·45	(6)=(3)−0·25(1)
	0·20	0·24	0·12	1	0	0		(7)=(4)
	0	0·12	0·18	−0·5	1	0		(8)=(5)
−2	0	0	0·10	0·75	−2	1	−0·15	(9)=(6)−2(5)
The inverse matrix $\mathbf{A}^{-1}$ is placed here ($\rightarrow$); its elements are calculated to $2S$ in the order shown (numbered 1, 2, ..., 9).	³ 19 ² −15 ¹ 7·5			⁶ −34 ⁵ 38 ⁴ −20				

We now consider an example of back-substitution calculations. $\widetilde{\mathbf{A}}$ taken with the 2nd column of $\widetilde{\mathbf{I}}$ yields the 2nd column of $\mathbf{A}^{-1}$. Thus:

$$\begin{bmatrix} 0·20 & 0·24 & 0·12 \\ 0 & 0·12 & 0·18 \\ 0 & 0 & 0·10 \end{bmatrix} \begin{bmatrix} u_6 \\ u_5 \\ u_4 \end{bmatrix} = \begin{bmatrix} 0 \\ 1 \\ -2 \end{bmatrix}$$

This yields $u_4 = -20$, $u_5 = 38$, and $u_6 = -34$, found in that order.

The reader should check by multiplication that $\mathbf{A}\mathbf{A}^{-1}$ is

$$\begin{bmatrix} 1·10 & -0·08 & 0·00 \\ 0·10 & 0·92 & 0·00 \\ 0·13 & -0·10 & 1·00 \end{bmatrix},$$

which is approximately equal to $\mathbf{I}$. The approximation is poor, due to the calculation of the elements of $\mathbf{A}^{-1}$ being made to $2S$ only.

3 SOLUTION OF LINEAR SYSTEMS, USING THE INVERSE MATRIX

As pointed out above, the solution to a linear system $\mathbf{A}\mathbf{x} = \mathbf{b}$ is $\mathbf{x} = \mathbf{A}^{-1}\mathbf{b}$. We shall illustrate this by solving an example which makes use of the matrix $\mathbf{A}^{-1}$ obtained in Section 2 above.

Example

We want to solve

$$0.20x + 0.24y + 0.12z = 1,$$

$$0.10x + 0.24y + 0.24z = 2,$$

$$0.05x + 0.30y + 0.49z = 3.$$

The solution is as follows. The coefficient matrix is

$$\mathbf{A} = \begin{bmatrix} 0.20 & 0.24 & 0.12 \\ 0.10 & 0.24 & 0.24 \\ 0.05 & 0.30 & 0.49 \end{bmatrix}$$

Assuming that the elements of $\mathbf{A}$ are known to $2S$ accuracy, we may compute $\mathbf{A}^{-1}$ as shown in Section 1 and use it as follows:

$$\mathbf{x} = \begin{bmatrix} x \\ y \\ z \end{bmatrix} = \mathbf{A}^{-1}\mathbf{b} = \begin{bmatrix} 19.0 & -34.0 & 12.0 \\ -15.0 & 38.0 & -15.0 \\ 7.5 & -20.0 & 10.0 \end{bmatrix} \begin{bmatrix} 1 \\ 2 \\ 3 \end{bmatrix} = \begin{bmatrix} -13.0 \\ 16.0 \\ -2.5 \end{bmatrix}.$$

That is, the solution to $2S$ is $x = -13.0$, $y = 16.0$, $z = -2.5$.

We may check the solution by adding the three equations. This yields

$$0.35x + 0.78y + 0.85z = 6.$$

Inserting the solution in the left hand side gives

$$0.35(-13) + 0.78 \times 16 + 0.85(-2.5) = 5.8 \text{ to } 2S.$$

Checkpoint

---

1. In the method for finding the inverse of $\mathbf{A}$, what is the final form of $\mathbf{A}$ after the elementary row operations have been carried out?
2. Is the solution of the system $\mathbf{Mx} = \mathbf{d}$, $\mathbf{x} = \mathbf{dM}^{-1}$ or $\mathbf{x} = \mathbf{M}^{-1}\mathbf{d}$ (or neither)?
3. Give a condition for a matrix not to have an inverse.

EXERCISES

1. Find the inverses of the following matrices, using the elimination and back-substitution method.

a) $\begin{bmatrix} 2 & 6 & 4 \\ 6 & 19 & 12 \\ 2 & 8 & 14 \end{bmatrix}$ b) $\begin{bmatrix} 1{\cdot}3 & 4{\cdot}6 & 3{\cdot}1 \\ 5{\cdot}6 & 5{\cdot}8 & 7{\cdot}9 \\ 4{\cdot}2 & 3{\cdot}2 & 4{\cdot}5 \end{bmatrix}$ c) $\begin{bmatrix} {\cdot}37 & {\cdot}65 & {\cdot}81 \\ {\cdot}41 & {\cdot}71 & {\cdot}34 \\ {\cdot}11 & {\cdot}82 & {\cdot}52 \end{bmatrix}$

2. Solve the following systems of equations (each with two right hand side vectors).

a) $\begin{aligned} 2x + 6y + 4z &= 5 \\ 6x + 19y + 12z &= 6 \\ 2x + 8y + 14z &= 7 \end{aligned}$ $\begin{bmatrix} = 1 \\ = 2 \\ = 3 \end{bmatrix}$

b) $\begin{aligned} 1{\cdot}3x + 4{\cdot}6y + 3{\cdot}1z &= -1 \\ 5{\cdot}6x + 5{\cdot}8y + 7{\cdot}9z &= 2 \\ 4{\cdot}2x + 3{\cdot}2y + 4{\cdot}5z &= -3 \end{aligned}$ $\begin{bmatrix} = 0 \\ = 1 \\ = 1 \end{bmatrix}$

c) $\begin{aligned} 0{\cdot}37x_1 + 0{\cdot}65x_2 + 0{\cdot}81x_3 &= 1{\cdot}1 \\ 0{\cdot}41x_1 + 0{\cdot}71x_2 + 0{\cdot}34x_3 &= 2{\cdot}2 \\ 0{\cdot}11x_1 + 0{\cdot}82x_2 + 0{\cdot}52x_3 &= -0{\cdot}1 \end{aligned}$ $\begin{bmatrix} = 0{\cdot}5 \\ = 2{\cdot}1 \\ = 1{\cdot}2 \end{bmatrix}$

FINITE DIFFERENCES 1
Tables

Traditionally, numerical analysts have been concerned with tables of numbers, and many techniques have been developed for dealing with mathematical functions represented in this way. For example, the value of the function at an untabulated point may be required, so that an *interpolation* procedure is necessary. It is also possible to estimate the *derivative* or the *definite integral* of a tabular function, using some finite processes to approximate the corresponding (infinitesimal) limiting procedures of calculus. In each case, it is traditional to use *finite differences*.

1 TABLES OF VALUES

Many books contain tables of mathematical functions. One of the most comprehensive is *Handbook of Mathematical Functions*, edited by M. Abramowitz and I. A. Stegun (U. S. National Bureau of Standards, 1964; Dover 1965), which also contains useful information about numerical methods.

Although most tables use constant argument intervals, some functions do change rapidly in value in particular regions of the argument, and hence may best be tabulated using intervals varying according to the local behaviour of the function. Tables with varying argument interval are much more difficult to work with, however, and it is common to adopt uniform argument intervals wherever possible. As a simple example consider the 6S table of the exponential function over 0·10 (0·01) 0·14 (this notation specifies the domain 0·1 $\leqslant x \leqslant$ 0·14 spanned in intervals of 0·01).

x	$f(x) = e^x$
0·10	1·10517
0·11	1·11628
0·12	1·12750
0·13	1·13883
0·14	1·15027

It is extremely important that the interval between successive values is small enough to display the variation of the tabulated function, because usually the value of the function will be needed at some argument value between values specified (e.g. e^x at $x = 0.105$ from the above table). If the table is so constructed, we can obtain such intermediate values of the function to good approximation by assuming a polynomial representation (hopefully, of low degree).

2 FINITE DIFFERENCES

Since Newton, *finite differences* have been used extensively. The construction of a table of finite differences for a tabular function is simple: *first differences* are obtained by subtracting each value from the succeeding value in a table, *second differences* by repeating this operation on the first differences, and so on for higher orders. For the above table of e^x: 0.10 (0.01) 0.14, one has the following table (note the customary layout, with decimal points and leading zeros omitted from the differences).

x	$f(x) = e^x$	First differences	Second differences	Third differences
0.10	1.10517			
		1111		
0.11	1.11628		11	
		1122		0
0.12	1.12750		11	
		1133		0
0.13	1.13883		11	
		1144		
0.14	1.15027			

(In this case, the differences must be multiplied by 10^{-5} for comparison with the function values.)

3 INFLUENCE OF ROUND-OFF ERRORS

Consider the difference table for $f(x) = e^x$: 0.1 (0.05) 0.5 to six significant figures constructed as in the case of the preceding example. As before, differences of increasing order decrease rapidly in magnitude, but the third differences are irregular. This is largely a consequence of round-off errors, as tabulation of the function to seven significant figures and differencing to *fourth* order illustrates (c.f. Exercises 2 below).

x	$f(x)=e^x$	First differences	Second differences	Third differences
0·1	1·10517			
		5666		
0·15	1·16183		291	
		5957		15
0·2	1·22140		306	
		6263		14
0·25	1·28403		320	
		6583		18
0·3	1·34986		338	
		6921		16
0·35	1·41907		354	
		7275		20
0·4	1·49182		374	
		7649		18
0·45	1·56831		392	
		8041		
0·5	1·64872			

Although the round-off errors in $f(x)$ should be less than $\frac{1}{2}$ in the last significant place, they may accumulate; the *greatest* error that can be obtained corresponds to:

Tabular error	Differences					
	1st.	2nd.	3rd.	4th.	5th.	6th.
+ 1/2						
	−1					
− 1/2		+2				
	+1		−4			
+ 1/2		−2		+8		
	−1		+4		−16	
− 1/2		+2		−8		+32
	+1		−4		+16	
+ 1/2		−2		+8		
	−1		+4			
− 1/2		+2				
	+1					
+ 1/2						

A rough working criterion for the *expected* fluctuations ('noise level') due to round-off error is shown in the following table.

Order of difference	1	2	3	4	5	6
Expected error limits	± 1	± 2	± 3	± 6	± 12	± 22

Checkpoint

---

1. What factors determine the intervals of tabulation of a function?
2. What is the name of the procedure to determine a value of a tabulated function at an intermediate point?
3. What may be the cause of irregularity in the highest order differences in a difference table?

EXERCISES

1. Construct a table of differences for the function $f(x) = x^3$ for $x = 0$ (1) 6.
2. Construct a difference table for the function $f(x) = e^x$ (given to 7 significant figures) for 0.1 (0.05) 0.5 :

x	$f(x)$	x	$f(x)$	x	$f(x)$
0·1	1·105171	0·25	1·284025	0·4	1·491825
0·15	1·161834	0·3	1·349859	0·45	1·568312
0·2	1·221403	0·35	1·419068	0·5	1·648721

FINITE DIFFERENCES 2
Forward, backward and central difference notations

There are three different notations for the single set of finite differences described in the previous Step; these define *forward*, *backward* and *central* differences. We introduce each of these three notations in terms of the so-called *shift operator*.

1 THE SHIFT OPERATOR E

Let $\{f_0, f_1, ..., f_{n-1}, f_n\}$ denote a set of values of the function $f(x)$ defined by $f_j \equiv f(x_j)$, where $x_j = x_0 + jh, j = 0, 1, ..., n$. The *shift operator* E is defined by

$$Ef_j = f_{j+1}.$$

Consequently,

$$E^2 f_j = E(Ef_j) = Ef_{j+1} = f_{j+2},$$

and so on; i.e.

$$E^k f_j = f_{j+k},$$

where k is any integer. Moreover, let us extend this last formula to all real values of j and k, so that for example

$$E^{\frac{1}{2}} f_j = f_{j+\frac{1}{2}} = f(x_0 + (j+\tfrac{1}{2})h).$$

and $\qquad E^{\frac{1}{2}} f_{j+\frac{1}{2}} = f_{j+1} = f(x_0 + (j+1)h).$

2 THE FORWARD DIFFERENCE OPERATOR Δ.

If we define the forward difference operator Δ by

$$\Delta = E - 1,$$

it follows that

$$\Delta f_j = (E-1)f_j = Ef_j - f_j = f_{j+1} - f_j,$$

which is the first order *forward difference* at x_j. Similarly,

$$\Delta^2 f_j = \Delta(\Delta f_j) = \Delta(f_{j+1} - f_j) = \Delta f_{j+1} - \Delta f_j = f_{j+2} - 2f_{j+1} + f_j$$

is the second order forward difference at x_j, and so on. The kth order forward difference is $\Delta^k f_j = \Delta^{k-1} f_{j+1} - \Delta^{k-1} f_j$,

where k is any integer.

3 THE BACKWARD DIFFERENCE OPERATOR ∇.

If we define the *backward difference operator* ∇ by

$$\nabla = 1 - E^{-1},$$

it follows that

$$\nabla f_j = (1 - E^{-1})f_j = f_j - E^{-1}f_j = f_j - f_{j-1},$$

which is the first order *backward difference* at x_j. Similarly,

$$\nabla^2 f_j = \nabla(\nabla f_j) = \nabla(f_j - f_{j-1}) = \nabla f_j - \nabla f_{j-1} = f_j - 2f_{j-1} + f_{j-2}$$

is the second order backward difference at x_j, and so on. The kth order backward difference is $\nabla^k f_j = \nabla^{k-1} f_j - \nabla^{k-1} f_{j-1}$,
where k is any integer.
Note that $\nabla f_j = \Delta f_{j-1}$ and $\nabla^k f_j = \Delta^k f_{j-k}$.

4 THE CENTRAL DIFFERENCE OPERATOR δ

If we define the *central difference operator* δ by

$$\delta = E^{\frac{1}{2}} - E^{-\frac{1}{2}},$$

it follows that

$$\delta f_j = (E^{\frac{1}{2}} - E^{-\frac{1}{2}})f_j = E^{\frac{1}{2}}f_j - E_j^{-\frac{1}{2}}f_j = f_{j+\frac{1}{2}} - f_{j-\frac{1}{2}}$$

which is the first order *central difference* at x_j. Similarly,

$$\delta^2 f_j = \delta(\delta f_j) = \delta(f_{j+\frac{1}{2}} - f_{j-\frac{1}{2}}) = f_{j+1} - 2f_j + f_{j-1}$$

is the second order central difference at x_j, and so on. The kth order central difference is $\delta^k f_j = \delta^{k-1} f_{j+\frac{1}{2}} - \delta^{k-1} f_{j-\frac{1}{2}}$,
where k is any integer.
Note that $\delta f_{j+\frac{1}{2}} = \Delta f_j = \nabla f_{j+1}$.

5 DIFFERENCE DISPLAY

The role of the forward, central and backward differences is displayed by the difference table:

x	$f(x)$	First difference	Second	Third	Fourth
x_0	f_0				
		Δf_0			
x_1	f_1		$\Delta^2 f_0$		
		Δf_1		$\Delta^3 f_0$	
x_2	f_2		$\Delta^2 f_1$		$\Delta^4 f_0$
		Δf_2		$\Delta^3 f_1$	
x_3	f_3		$\Delta^2 f_2$		
		Δf_3			
x_4	f_4				
.	.				
.	.				
.	.				
.	.				
x_{j-2}	f_{j-2}				
		$\delta f_{j-3/2}$			
x_{j-1}	f_{j-1}		$\delta^2 f_{j-1}$		
		$\delta f_{j-1/2}$		$\delta^3 f_{j-1/2}$	
x_j	f_j		$\delta^2 f_j$		$\delta^4 f_j$
		$\delta f_{j+1/2}$		$\delta^3 f_{j+1/2}$	
x_{j+1}	f_{j+1}		$\delta^2 f_{j+1}$		
		$\delta f_{j+3/2}$			
x_{j+2}	f_{j+2}				
.	.				
.	.				
.	.				
.	.				
x_{n-4}	f_{n-4}				
		∇f_{n-3}			
x_{n-3}	f_{n-3}		$\nabla^2 f_{n-2}$		
		∇f_{n-2}		$\nabla^3 f_{n-1}$	
x_{n-2}	f_{n-2}		$\nabla^2 f_{n-1}$		$\nabla^4 f_n$
		∇f_{n-1}		$\nabla^3 f_n$	
x_{n-1}	f_{n-1}		$\nabla^2 f_n$		
		∇f_n			
x_n	f_n				

Although forward, central and backward differences represent precisely the same set of numbers,

i)   forward differences are especially useful near the start of a table, since they involve tabulated function values *below* x_j;

ii) central differences are especially useful away from the ends of the table, where there are available tabulated function values *above and below* x_j;

iii) backward differences are especially useful near the end of a table, since they involve tabulated function values *above* x_j.

Checkpoint

---

1. What is the definition of the shift operator?
2. How are the forward, backward and central difference operators defined?
3. When are the respective forward, backward and central difference notations likely to be used?

EXERCISES

1. Reconsider the difference table (in Section 3 of Step 15) of $f(x) = e^x$ for $x = 0.1 \, (0.05) \, 0.5$ to five significant figures, and determine the following (taking $x_0 = 0.1$):

    i) $\Delta f_2, \quad \Delta^2 f_2, \quad \Delta^3 f_2, \quad \Delta^4 f_2$;

    ii) $\nabla f_6, \quad \nabla^2 f_6, \quad \nabla^3 f_6, \quad \nabla^4 f_6$;

    iii) $\delta^2 f_4, \quad \delta^4 f_4$;

    iv) $\Delta^2 f_1, \quad \delta^2 f_2, \quad \nabla^2 f_3$;

    v) $\Delta^3 f_3, \quad \nabla^3 f_6, \quad \delta^3 f_{\frac{9}{2}}$.

2. Prove that

    i) $E x_j = x_{j+1}$;

    ii) $\Delta^3 f_j = f_{j+3} - 3f_{j+2} + 3f_{j+1} - f_j$;

    iii) $\nabla^3 f_j = f_j - 3f_{j-1} + 3f_{j-2} - f_{j-3}$;

    iv) $\delta^3 f_j = f_{j+\frac{3}{2}} - 3f_{j+\frac{1}{2}} + 3f_{j-\frac{1}{2}} - f_{j-\frac{3}{2}}$;

FINITE DIFFERENCES 3
Polynomials

Since polynomial approximations are used in many areas of Numerical Analysis, it is important to investigate the effects of differencing polynomials.

1 FINITE DIFFERENCES OF A POLYNOMIAL

Consider the finite differences of an nth degree polynomial

$$f(x) = a_n x^n + a_{n-1} x^{n-1} + \ldots + a_1 x + a_0$$

tabulated for equidistant points at tabular interval h.

THEOREM: The nth difference of a polynomial of degree n is a constant proportional to h^n, and higher order differences are zero.

Proof: Omitting the subscript on x_j, we have

$$\Delta f(x) = f(x+h) - f(x)$$
$$= a_n[(x+h)^n - x^n] + a_{n-1}[(x+h)^{n-1} - x^{n-1}] + \ldots + a_1[(x+h) - x]$$
$$= a_n n x^{n-1} h + \text{polynomial of degree } n - 2.$$

$$\Delta^2 f(x) = a_n n h[(x+h)^{n-1} - x^{n-1}] + \ldots$$
$$= a_n n(n-1) x^{n-2} h^2 + \text{polynomial of degree } n - 3.$$

$$\cdot \quad \cdot \quad \cdot \quad \cdot \quad \cdot \quad \cdot \quad \cdot \quad \cdot \quad \cdot \quad \cdot \quad \cdot \quad \cdot \quad \cdot \quad \cdot \quad \cdot \quad \cdot \quad \cdot$$

$$\Delta^n f(x) = a_n n! h^n = \text{constant}$$

$$\Delta^{n+1} f(x) = 0.$$

In passing, the student may recall that in differential calculus the *increment* $\Delta f(x) = f(x + h) - f(x)$ is related to the derivative of $f(x)$ at the point x.

Example

$$f(x) = x^3 \text{ for } x = 5\cdot0 \ (0\cdot1) \ 5\cdot5.$$

x	$f(x) = x^3$	Δ	Δ^2	Δ^3	Δ^4
5·0	125·000				
		7651			
5·1	132·651		306		
		7957		6	
5·2	140·608		312		0
		8269		6	
5·3	148·877		318		0
		8587		6	
5·4	157·464		324		
		8911			
5·5	166·375				

In this case $a_n = 1, n = 3, h = 0\cdot1$, whence $\Delta^3 f(x) = 1 \times 3! \times (0\cdot1)^3$
$= 0\cdot006$.

Note that round-off error noise may occur : e.g., consider the tabulation of $f(x) = x^3$ for 5·0 (0·1) 5·5 rounded to two decimal places.

x	$f(x) = x^3$	Δ	Δ^2	Δ^3	Δ^4
5·0	125·00				
		765			
5·1	132·65		31		
		796		0	
5·2	140·61		31		0
		827		0	
5·3	148·88		31		3
		858		3	
5·4	157·46		34		
		892			
5·5	166·38				

2 APPROXIMATION OF A FUNCTION BY A POLYNOMIAL

Whenever the higher differences of a table become small (allowing for round-off noise), the function represented may be well approximated by a polynomial. For example, reconsider the difference table of $f(x) = e^x$ for $x = 0\cdot1 \ (0\cdot05) \ 0\cdot5$ to six significant figures.

x	$f(x) = e^x$	Δ	Δ^2	Δ^3	Δ^4
0·1	1·10517				
		5666			
0·15	1·16183		291		
		5957		15	
0·2	1·22140		306		−1
		6263		14	
0·25	1·28403		320		4
		6583		18	
0·3	1·34986		338		−2
		6921		16	
0·35	1·41907		354		4
		7275		20	
0·4	1·49182		374		−2
		7649		18	
0·45	1·56831		392		
		8041			
0·5	1·64872				

Since the estimate for round-off error at Δ^3 is $\pm$ 3 (see p. 71), we say that third differences are constant within round-off error, and deduce that a *cubic* approximation is appropriate for e^x over the range 0·1 $< x < 0·5$ at interval 0·05. In this fashion, differences can be used to decide what (if any) degree of approximating polynomial is appropriate.

An example in which polynomial approximation is inappropriate is $f(x) = 10^x$ for $x = 0\,(1)\,4$, thus

x	$f(x)$	Δ	Δ^2	Δ^3	Δ^4
0	1				
		9			
1	10		81		
		90		729	
2	100		810		6561
		900		7290	
3	1000		8100		
		9000			
4	10000				

Although $f(x) = 10^x$ is 'smooth', the large tabular interval ($h = 1$) produces large higher order finite differences. It should also be understood that there exist functions that cannot usefully be tabulated at all, at least in some neighbourhood; e.g., $f(x) = \sin(1/x)$ near the origin $x = 0$. Nevertheless, these are fairly exceptional cases.

Finally, we remark that the approximation of a function by a polynomial is fundamental to the widespread use of finite difference methods.

Checkpoint

1. What may be said about the higher order (exact) differences of a polynomial?
2. What is the effect of round-off error on the higher order differences of a polynomial?
3. When may a function be approximated by a polynomial?

EXERCISES

1. Construct a difference table for the polynomial $f(x) = x^4$ for $x = 0\,(0\cdot1)\,1$ when
   i) the values of $f(x)$ are exact;
   ii) the values of $f(x)$ have been rounded to 3 decimals.
   Compare the fourth difference round-off errors with the estimate $\pm\,6$.
2. Find the degree of the polynomial which fits the data in the following table.

x	$f(x)$	x	$f(x)$
0	3	3	24
1	2	4	59
2	7	5	118

FINITE DIFFERENCES 4
Detection and correction of mistakes

We have seen that round-off errors cause small fluctuations in differences. Mistakes produce much larger fluctuations, so that a powerful check of tabulated functions is to obtain the finite differences. It is recommended that this check be applied to any table produced by calculation.

1 THE FAN DUE TO A MISTAKE

An error ε in the value of $f(x)$ or in a difference propagates as follows:

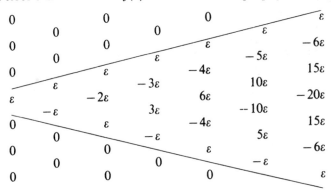

A pair of adjacent errors propagates as follows:

In each case there is an 'error fan' which indicates the position of the mistake. The coefficients in the first fan form a familiar pattern: apart from the minus signs we have Pascal's triangle, i.e. the binomial coefficients.

In practice however, one may locate and correct mistakes in tables. If the function has been correctly tabulated, at such an interval that it is well-represented by a polynomial, (eventually) we expect to find a column of approximately constant differences. If instead we find a column containing a sequence of differences with alternating sign that are significantly bigger than other differences in the same column, we suspect one or more mistakes, and use our knowledge of the error fans to locate and correct them (c.f. Examples).

2 EXAMPLES

i) In the tabulation below of $f(x) = x^3$ for $x = 5{\cdot}0\ (0{\cdot}1)\ 6{\cdot}0$, a mistake at $x = 5{\cdot}5$ is indicated by the error fan:

x	$f(x) = x^3$	Δ	Δ^2	Δ^3	Δ^4
5·0	125·000				
		7651			
5·1	132·651		306		
		7957		6	
5·2	140·608		312		0
		8269		6	
5·3	148·877		318		−18
		8587		−12	
5·4	157·464		306		72
		8893		60	
5·5	166·357		366		−108
		9259		−48	
5·6	175·616		318		72
		9577		24	
5·7	185·193		342		−18
		9919		6	
5·8	195·112		348		0
		10267		6	
5·9	205·379		354		
		10621			
6·0	216·000				

The mistake is shown up by fluctuations in Δ^3 and Δ^4. Since the function is cubic we can either compare the sequences $\{-18,\ 72,\ -108,\ 72,\ -18\}$ and $\{\varepsilon,\ -4\varepsilon,\ 6\varepsilon,\ -4\varepsilon,\ \varepsilon\}$ or compare the sequences $\{-12,\ 60,$

-48, $24\}$ and $\{6+\varepsilon, 6-3\varepsilon, 6+3\varepsilon, 6-\varepsilon\}$. Either way we deduce that $\varepsilon = -18$ and $f(5\cdot5) = 166\cdot375$. The mistake involved transposing adjacent digits (see Step 2).

Suppose we now correct the transposition mistake, but make another mistake in the consequent difference table, thus:

x	$f(x) = x^3$	Δ	Δ^2	Δ^3	Δ^4
5·0	125·000				
		7651			
5·1	132·651		306		
		7957		6	
5·2	140·608		312		0
		8269		6	
5·3	148·877		318		0
		8587		6	
5·4	157·464		324		-27
		8911		-21	
5·5	166·375		303		81
		9214		60	
5·6	175·616		363		-81
		9577		-21	
5·7	185·193		342		27
		9919		6	
5·8	195·112		348		0
		10267		6	
5·9	205·379		354		
		10621			
6·0	216·000				

Once again, the fan indicates the mistake of magnitude $\varepsilon = -27$ (c.f. sequence $\{-27, 81, -81, 27\}$), due to the entry 9214 rather than the correct entry 9241. The student may like to complete the correct difference table.

ii) In the following tabulation (see p. 84) of $f(x) = \log_{10} x$ for $x = 1\cdot0$ (0·1) 2·4, *two* mistakes are detected and corrected.

The fluctuations in Δ^2 make us suspicious, and our suspicions are confirmed by the behaviour of Δ^3. The sequence of differences $\{-54, 183, -178, 93, -87, 90, -29\}$ has alternating sign and all members are much bigger than the other third differences. We expect to find the pattern $\varepsilon, -3\varepsilon, 3\varepsilon, -\varepsilon$ so we deduce that there are mistakes in $f(1\cdot4)$ and $f(1\cdot7)$.

If we suppose that $0\cdot1401 = f(1\cdot4) + \varepsilon_1$ and $0\cdot2334 = f(1\cdot7) + \varepsilon_2$, we would expect the propagated errors in Δ^3 to be $\{\varepsilon_1, -3\varepsilon_1, 3\varepsilon_1,$

x	$f(x) = \log_{10}x$	Δ	Δ^2	Δ^3	Δ^3 after first correction	Δ^3 after second correction
1·0	0·0000					
		414				
1·1	0·0414		−36			
		378		5	5	5
1·2	0·0792		−31			
		347		−54	6	6
1·3	0·1139		−85			
		262		183	3	3
1·4	0·1401		+98			
		360		−178	2	2
1·5	0·1761		−80			
		280		93	33	3
1·6	0·2041		13			
		293		−87	−87	3
1·7	0·2334		−74			
		219		90	90	0
1·8	0·2553		16			
		235		−29	−29	1
1·9	0·2788		−13			
		222		3	3	3
2·0	0·3010		−10			
		212		0	0	0
2·1	0·3222		−10			
		202		1	1	1
2·2	0·3424		−9			
		193		1	1	1
2·3	0·3617		−8			
		185				
2·4	0·3802					

$-\varepsilon_1 + \varepsilon_2, -3\varepsilon_2, 3\varepsilon_2, -\varepsilon_2\}$. Assuming that third differences are approximately constant, we can use $183 \approx c - 3\varepsilon_1$ and $-178 \approx c + 3\varepsilon_1$ to estimate $\varepsilon_1 \approx \dfrac{-183 - 178}{6} \approx -60$. The results of making this correction are shown above. Similarly $-87 \approx c - 3\varepsilon_2$ and $90 \approx c + 3\varepsilon_2$ suggest that $\varepsilon_2 \approx \dfrac{87 + 90}{6} \approx 30$.

The *suggested* correct values are therefore obtained from

$$f(1·4) + \varepsilon_1 = 0·1401$$
$$f(1·7) + \varepsilon_2 = 0·2334$$

viz,

$$f(1\cdot4) = 0\cdot1401 + 0\cdot0060 = 0\cdot1461$$
$$f(1\cdot7) = 0\cdot2334 - 0\cdot0030 = 0\cdot2304$$

After the second correction has been made, the third differences look much more reasonable. Note that when estimating ε we prefer to use larger entries in the relevant column in order to minimize the effect of genuine variation in the differences due to round-off.

Checkpoint

1. How may mistakes in tables be found?
2. What check should be made on a function tabulated by calculation?
3. How are mistakes in tables corrected?

EXERCISES

1. The following table of exact values of a cubic $f(x)$ contains a mistake. Find the wrong entry and correct it.

x	$f(x)$	x	$f(x)$	x	$f(x)$	x	$f(x)$
2	3·0671	5	13·3184	8	24·2573	11	35·9000
3	6·4088	6	16·8875	9	28·0592	12	39·9401
4	9·8257	7	20·5366	10	31·9399	13	44·0608

2. Find and correct the mistakes in the following table.

x	$f(x)$	x	$f(x)$	x	$f(x)$
0	1·3246	4	1·6332	8	1·9282
1	1·4031	5	1·7082	9	2·0000
2	1·4807	6	1·7823	10	2·0711
3	1·5547	7	1·8577	11	2·1414

INTERPOLATION 1
Linear and quadratic interpolation

Interpolation is 'the art of reading between the lines in a table' and may be regarded as a special case of the general process of *curve fitting* (see Step 25). More precisely, interpolation is the process whereby non-tabulated values of a tabular function are estimated, on the assumption that the function behaves sufficiently smoothly between tabular points for it to be approximated by a polynomial of fairly low degree.

Interpolation is not as important in Numerical Analysis as it was, now that automatic computers (and electronic desk calculators with function keys) are available, and function values may often be obtained readily by algorithm (probably from a standard subroutine). However

i)   interpolation is important for functions known only as tables; and
ii) interpolation serves to introduce the wider application of finite differences.

In Step 17, we observed that when kth order differences are constant (within round-off fluctuation), the tabular function may be approximated by a polynomial of degree k. *Linear* and *quadratic* interpolation correspond to cases $k = 1$ and $k = 2$, respectively.

1 LINEAR INTERPOLATION

When a function varies so slowly that first differences are approximately constant, it may be approximated closely by a straight line between adjacent tabular points. This is the basic idea of *linear interpolation*, familiar from common use in tables of logarithms and trigonometric functions.

In Figure 10, the two function points (x_j, f_j) and (x_{j+1}, f_{j+1}) are connected by a straight line. Any x between x_j and x_{j+1} may be defined by a value θ such that

$$x - x_j = \theta(x_{j+1} - x_j) \equiv \theta h, \qquad 0 < \theta < 1.$$

Provided $f(x)$ is only slowly varying in the interval, the value of the

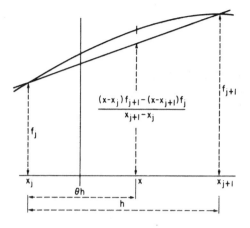

FIGURE 10. Linear interpolation

function at x is approximately given by the ordinate to the straight line at x. Elementary geometrical considerations yield

$$\theta = \frac{x - x_i}{x_{j+1} - x_j} \approx \frac{f(x) - f_j}{f_{j+1} - f_j},$$

so that

$$
\begin{aligned}
f(x) &\approx f_j + \theta(f_{j+1} - f_j) \\
&= f_j + \theta \Delta f_j \\
&= f_j + \theta \nabla f_{j+1} \\
&= f_j + \theta \delta f_{j+\frac{1}{2}}.
\end{aligned}
$$

In analytical terms, we have approximated $f(x)$ by

$$P_1(x) = f_j + \frac{x - x_j}{x_{j+1} - x_j}(f_{j+1} - f_j),$$

a linear function of x which satisfies

$$P_1(x_j) = f_j = f(x_j), \qquad P_1(x_{j+1}) = f_{j+1} = f(x_{j+1}).$$

Example

From a 4D table of e^{-x} we obtain the following difference table. The first differences are almost constant locally, so that the table is

x	$f(x)$	Δ	Δ^2
0·90	0·4066		
		-41	
0·91	0·4025		1
		-40	
0·92	0·3985		1
		-39	
0·93	0·3946		-1
		-40	
0·94	0·3906		1
		-39	
0·95	0·3867		1
		-38	
0·96	0·3829		0
		-38	
0·97	0·3791		0
		-38	
0·98	0·3753		1
		-37	
0·99	0·3716		

suitable for linear interpolation. For example,

$$f(0{\cdot}934) = 0{\cdot}3946 + \frac{4}{10}(-0{\cdot}0040) = 0{\cdot}3930$$

2   QUADRATIC INTERPOLATION

As remarked, linear interpolation is appropriate only for closely tabulated, slowly varying functions. The next simplest process is *quadratic interpolation*, based on an approximating polynomial of degree 2; one might expect that this approximation would give better accuracy for functions with larger variation.

Given three adjacent points x_j, $x_{j+1} = x_j + h$ and $x_{j+2} = x_j + 2h$, suppose that $f(x)$ is approximated by

$$P_2(x) = a + b(x - x_j) + c(x - x_j)(x - x_{j+1}),$$

where a, b, and c are chosen so that

$$P_2(x_{j+k}) = f(x_{j+k}) = f_{j+k}, \qquad k = 0, 1, 2.$$

Thus
$$P_2(x_j)\ \ = a = f_j$$
$$P_2(x_{j+1}) = a + bh = f_{j+1}$$
$$P_2(x_{j+2}) = a + 2bh + 2ch^2 = f_{j+2},$$

whence

$$a = f_j$$
$$b = (f_{j+1} - a)/h = (f_{j+1} - f_j)/h = \Delta f_j/h$$
$$c = (f_{j+2} - 2bh - a)/(2h^2) = (f_{j+2} - 2f_{j+1} + f_j)/(2h^2)$$
$$= \Delta^2 f_j/(2h^2).$$

Setting $x = x_j + \theta h$, we obtain the quadratic interpolation formula

$$f(x_j + \theta h) \approx f_j + \theta \Delta f_j + \tfrac{1}{2}\theta(\theta - 1)\Delta^2 f_j.$$

We note immediately that the quadratic interpolation formula introduces a second order term (involving $\Delta^2 f_j$) not included in the linear interpolation formula.

Example

Determine the second order correction to the value of $f(0\cdot934)$ obtained above using linear interpolation.

The second order correction is

$$\frac{1}{2} \times \frac{4}{10} \times \left(-\frac{6}{10}\right)(+0\cdot0001) = -\frac{0\cdot0024}{200},$$

so that the quadratic interpolation formula gives

$$f(0\cdot934) = 0\cdot3946 - \frac{0\cdot0160}{10} - \frac{0\cdot0024}{200} = 0\cdot3930.$$

The correction $-\dfrac{0\cdot0024}{200}$ is negligible.

Checkpoint

---

1. What is the process of obtaining a non-tabulated value of a function called?
2. When is linear interpolation adequate?
3. When is quadratic interpolation needed and adequate?

EXERCISES

1. Entries in a table of cos x are:

	0′	10′	20′	30′	40′	50′
80°	0·1736	0·1708	0·1679	0·1650	0·1622	0·1593

Obtain the value cos 80° 35′ from (i) linear interpolation, and (ii) quadratic interpolation.

2. Entries in a table of tan x are:

	0′	10′	20′	30′	40′	50′
80°	5·671	5·769	5·871	5·976	6·084	6·197

Determine whether linear or quadratic interpolation may be used. If so, obtain the value tan 80° 35′.

INTERPOLATION 2
Newton interpolation formulae

The linear and quadratic interpolation formulae are based on first and second degree polynomial approximation. Newton derived general forward and backward difference interpolation formulae for tables of constant interval h (corresponding to approximation by a polynomial of degree n).

1 NEWTON'S FORWARD DIFFERENCE FORMULA

Consider the points $x_j, x_j + h, x_j + 2h, \ldots$, and recall that

$$Ef_j = f_{j+1} = f(x_j + h), \qquad E^\theta f_j = f_{j+\theta} = f(x_j + \theta h),$$

where θ is any real number.

Formally, one has (since $\Delta = E - 1$)

$$f(x_j + \theta h) = E^\theta f_j$$

$$= (1 + \Delta)^\theta f_j$$

$$= [1 + \theta\Delta + \tfrac{1}{2}\theta(\theta-1)\Delta^2 + \frac{\theta(\theta-1)(\theta-2)}{3!}\Delta^3 + \ldots]f_j,$$

which is *Newton's forward difference formula*. The linear and quadratic (forward) interpolation formulae correspond to truncation at first and second order, respectively. If we truncate at nth order, we obtain

$$f(x_j + \theta h) \approx [1 + \theta\Delta + \tfrac{1}{2}\theta(\theta-1)\Delta^2 + \ldots + \frac{\theta(\theta-1)\ldots(\theta-n+1)}{n!}\Delta^n]f_j,$$

which is an approximation based on the values $f_j, f_{j+1}, \ldots, f_{j+n}$. It will be exact if (within round-off error)

$$\Delta^{n+k}f_j = 0, \qquad k = 1, 2, \ldots,$$

which is the case if $f(x)$ is a polynomial of degree n.

2   NEWTON'S BACKWARD DIFFERENCE FORMULA

Formally, one has (since $\nabla = 1 - E^{-1}$)

$$f(x_j + \theta h) = E^\theta f_j$$

$$= (1 - \nabla)^{-\theta} f_j$$

$$= (1 + \theta\nabla + \tfrac{1}{2}\theta(\theta+1)\nabla^2 + \frac{\theta(\theta+1)(\theta+2)}{3!}\nabla^3 + \ldots)f_j,$$

which is *Newton's backward difference formula*. The linear and quadratic (backward) interpolation formulae correspond to truncation at first or second order, respectively. The approximation based on the values $f_j, f_{j-1}, \ldots, f_{j-n}$ is

$$f(x_j + \theta h) \approx \left[1 + \theta\nabla + \tfrac{1}{2}\theta(\theta+1)\nabla^2 + \ldots + \frac{\theta(\theta+1)\ldots(\theta+n-1)}{n!}\nabla^n\right]f_j.$$

3   USE OF NEWTON'S INTERPOLATION FORMULAE

The Newton forward and backward difference formulae are well suited for use *at the beginning and end of a difference table*, respectively. (Other formulae that make use of central differences may be more convenient elsewhere.)

As an example, consider the following difference table of $f(x) = \sin x$ for $x = 0° (10°) 50°$.

$x°$	$f(x) = \sin x$	Δ	Δ^2	Δ^3	Δ^4	Δ^5
0	0·0000					
		1736				
10	0·1736		−52			
		1684		−52		
20	0·3420		−104		4	
		1580		−48		0
30	0·5000		−152		4	
		1428		−44		
40	0·6428		−196			
		1232				
50	0·7660					

   Since constant differences occur at fourth order, we conclude that a *quartic* approximation is appropriate. (Third order differences are not quite constant within expected round-off, and we anticipate that a cubic approximation is not quite good enough.) To determine $\sin 5°$ from the table, we use Newton's forward difference formula (to 4th order); thus

taking $x_j = 0$, we have $\theta = \dfrac{5-0}{10} = \frac{1}{2}$ $(h = 10)$, and

$$\sin 5° = \sin 0° + \tfrac{1}{2}(0{\cdot}1736) + \tfrac{1}{2}\tfrac{1}{2}\left(-\tfrac{1}{2}\right)(-0{\cdot}0052)$$

$$+ \tfrac{1}{6}\tfrac{1}{2}\left(-\tfrac{1}{2}\right)\left(-\tfrac{3}{2}\right)(-0{\cdot}0052) + \tfrac{1}{24}\tfrac{1}{2}\left(-\tfrac{1}{2}\right)\left(-\tfrac{3}{2}\right)\left(-\tfrac{5}{2}\right)(0{\cdot}0004)$$

$$= 0 + 0{\cdot}0868 + 0{\cdot}0006(5) - 0{\cdot}0003(3) - 0{\cdot}0000(2)$$

$$= 0{\cdot}0871 \quad \text{(c.f. } 0{\cdot}0872 \text{ from tables).}$$

Note that we have kept a guard digit (in brackets) to minimize accumulated round-off error. To determine $\sin 45°$ from the table, we use Newton's backward difference formula (to 4th order); thus taking $x_j = 40$, we have $\theta = \dfrac{45-40}{10} = \dfrac{1}{2}$, and

$$\sin 45° = \sin 40° + \tfrac{1}{2}(0{\cdot}1428) + \tfrac{1}{2}\tfrac{1}{2}\tfrac{3}{2}(-0{\cdot}0152)$$

$$+ \tfrac{1}{6}\tfrac{1}{2}\tfrac{3}{2}\tfrac{5}{2}(-0{\cdot}0048) + \tfrac{1}{24}\tfrac{1}{2}\tfrac{3}{2}\tfrac{5}{2}\tfrac{7}{2}(0{\cdot}0004)$$

$$= 0{\cdot}6428 + 0{\cdot}0714 - 0{\cdot}0057 - 0{\cdot}0015 + 0{\cdot}0001(1)$$

$$= 0{\cdot}7071 \quad \text{(c.f. } 0{\cdot}7071 \text{ from tables).}$$

4   UNIQUENESS OF THE INTERPOLATING POLYNOMIAL

Given a set of values $f(x_0), f(x_1), ..., f(x_n)$ with $x_j = x_0 + jh$, we have two nth order interpolation formulae available:

$$f(x) \approx P_n(x)$$
$$= (1 + \theta\Delta + \tfrac{1}{2}\theta(\theta-1)\Delta^2 + ... + \frac{\theta(\theta-1)...(\theta-n+1)}{n!}\Delta^n)f_0$$

and

$$f(x) \approx Q_n(x)$$
$$= (1 + \phi\nabla + \tfrac{1}{2}\phi(\phi+1)\nabla^2 + ... + \frac{\phi(\phi+1)...(\phi+n-1)}{n!}\nabla^n)f_n$$

where $\theta = (x-x_0)/h$ and $\phi = (x-x_n)/h$.

Clearly $P_n(x)$ and $Q_n(x)$ are both polynomials in x of degree n. It can be verified (see Exercise 2 below) that $P_n(x_j) = Q_n(x_j) = f(x_j)$

for $j = 0, 1, \ldots, n$, which implies that $P_n(x) - Q_n(x)$ is a polynomial of degree n which vanishes at $(n+1)$ points. This in turn implies that $P_n(x) - Q_n(x) \equiv 0$, or $P_n(x) \equiv Q_n(x)$. In fact a polynomial of degree n through any given $(n+1)$ (distinct but not necessarily equidistant) points is unique, and is called *the collocation polynomial*.

5   ANALOGY WITH TAYLOR SERIES

If we define for integer k

$$D^k f_j \equiv \frac{d^k f}{dx^k}\bigg|_{x = x_j}$$

the Taylor Series about x_j becomes

$$f(x) = f_j + (x - x_j) Df_j + \frac{(x - x_j)^2}{2!} D^2 f_j + \ldots .$$

Setting $x = x_j + \theta h$, we have formally

$$f(x_j + \theta h) = f_j + \theta h \, Df_j + \frac{\theta^2 h^2}{2!} D^2 f_j + \ldots$$

$$= \left[1 + \theta h \, D + \frac{\theta^2 h^2}{2!} D^2 + \ldots\right] f_j$$

$$= e^{\theta h D} f_j.$$

Comparison with the Newton interpolation form

$$f(x_j + \theta h) = E^\theta f_j,$$

shows that the operator e^{hD} (on functions of a continuous variable) is analogous to the operator E (on functions of a discrete variable).

Checkpoint

---

1. What is the relationship between the forward and backward linear and quadratic interpolation formulae (for a table of constant interval h) and Newton's interpolation formulae?
2. When is the Newton forward difference formula convenient to use?
3. When is the Newton backward difference formula convenient to use?

EXERCISES

1. From a difference table of $f(x) = e^x$ (using 5 decimals) for $x = 0.10 \, (0.05) \, 0.40$, estimate
   i) $e^{0.14}$ using Newton's forward difference formula; and
   ii) $e^{0.315}$ using Newton's backward difference formula.

2. Show that

$$f_j = f(x_0 + jh) = (1 + j\Delta + j \, (\frac{j-1}{2})\Delta^2 + \ldots + \Delta^j)f(x_0) \quad \text{for } j = 0, 1, 2, .$$

3. Derive the equation of the collocation polynomial for the following data.

x	$f(x)$	x	$f(x)$
0	3	3	24
1	2	4	59
2	7	5	118

INTERPOLATION 3*
Other interpolation formulae involving finite differences

Although any interpolation from tables of constant interval may be done using Newton's forward formula, there are other interpolation formulae which may be preferred. In the previous Step 20, we noted that Newton's backward formula might be preferred near the end of a difference table. Stirling, Bessel and Everett formulae are others suitable for use centrally across a difference table, and these formulae are considered in this optional Step. We first consider the Gauss interpolation formulae, however, from which the others readily follow.

At this stage, we emphasize that all of these formulae are based on the collocation polynomial; and we have already noted that the (at most) nth degree polynomial through any given subset of $(n+1)$ tabular points is unique. The choice of formula is determined merely by computational efficiency.

1 GAUSS' INTERPOLATION FORMULAE

Since

$$\Delta^2 f_j = \Delta^2(f_{j-1}+\Delta f_{j-1}) = \Delta^2 f_{j-1} + \Delta^3 f_{j-1}$$
$$\Delta^3 f_j = \Delta^3 f_{j-1} + \Delta^4(f_{j-2}+\Delta f_{j-2}) = \Delta^3 f_{j-1} + \Delta^4 f_{j-2} + \Delta^5 f_{j-2},$$
$$\text{etc.}$$

from Newton's forward formula, one has immediately

$$f(x_j+\theta h) = f_j + \theta\Delta f_j + \frac{\theta(\theta-1)}{2!}\Delta^2 f_{j-1} + \frac{(\theta+1)\theta(\theta-1)}{3!}\Delta^3 f_{j-1} + \dots,$$

which is *Gauss' forward formula*. Similarly, from

$$\Delta f_j = \Delta(f_{j-1}+\Delta f_{j-1}) = \Delta f_{j-1} + \Delta^2 f_{j-1}$$
$$\Delta^2 f_j = \Delta^2 f_{j-1} + \Delta^3(f_{j-2}+\Delta f_{j-2}) = \Delta^2 f_{j-1} + \Delta^3 f_{j-2} + \Delta^4 f_{j-2},$$
$$\text{etc.,}$$

and Newton's formula, one has

$$f(x_j + \theta h) = f_j + \theta \Delta f_{j-1} + \frac{(\theta+1)\theta}{2!} \Delta^2 f_{j-1} + \frac{(\theta+1)\theta(\theta-1)}{3!} \Delta^3 f_{j-2} + ...,$$

which is *Gauss' backward formula.*

It is clear from a difference table that these formulae use differences across the table:

—— Gauss' forward formula   – – – Gauss' backward formula.

This table also serves to indicate that it is common to write the Gauss formulae in central difference notation:

$$f(x_j + \theta h) = f_j + \theta \delta f_{j+\frac{1}{2}} + \frac{\theta(\theta-1)}{2!} \delta^2 f_j + \frac{(\theta+1)\theta(\theta-1)}{3!} \delta^3 f_{j+\frac{1}{2}} + ...,$$

$$f(x_j + \theta h) = f_j + \theta \delta f_{j-\frac{1}{2}} + \frac{(\theta+1)\theta}{2!} \delta^2 f_j + \frac{(\theta+1)\theta(\theta-1)}{3!} \delta^3 f_{j-\frac{1}{2}} + ...$$

The Gauss formulae are mainly of theoretical interest, and we now proceed to derive the other formulae mentioned above. We continue to follow common practice and use central difference notation.

2 STIRLING'S FORMULA

Adding the two Gauss formulae, and dividing by 2 throughout, immediately produces *Stirling's formula:*

$$f(x_j + \theta h) = f_j + \theta \mu \delta f_j + \frac{\theta^2}{2!} \delta^2 f_j + \frac{(\theta+1)\theta(\theta-1)}{3!} \mu \delta^3 f_j + ...,$$

where μ is the *mean value operator.* (In terms of the shift operator, $\mu = \frac{1}{2}(E^{\frac{1}{2}} + E^{-\frac{1}{2}})$. Thus, for example, one has $2\mu\delta f_j = \delta f_{j+\frac{1}{2}} + \delta f_{j-\frac{1}{2}}$ $= f_{j+1} - f_{j-1}$.)

Stirling's formula is convenient to use for small θ, say $|\theta| \leqslant \frac{1}{4}$.

3 BESSEL'S FORMULA

Since $x_j + \theta h = x_{j+1} - h + \theta h = x_{j+1} + (\theta - 1)h$,
one may set $(j+1)$ for j and $(\theta - 1)$ for θ, respectively, in Gauss' backward formula to get $f(x_j + \theta h) \equiv f(x_{j+1} + (\theta - 1)h)$

$$= f_{j+1} + (\theta - 1)\delta f_{j+\frac{1}{2}} + \frac{\theta(\theta - 1)}{2!} \delta^2 f_{j+1} + \frac{\theta(\theta - 1)(\theta - 2)}{3!} \delta^3 f_{j+\frac{1}{2}} + \dots .$$

Adding this result to Gauss' forward formula produces (after dividing throughout by 2)

$$f(x_j + \theta h) = \mu f_{j+\frac{1}{2}} + (\theta - \tfrac{1}{2})\delta f_{j+\frac{1}{2}} + \frac{\theta(\theta - 1)}{2!} \mu \delta^2 f_{j+\frac{1}{2}} +$$

$$+ \frac{\theta(\theta - 1)(\theta - \frac{1}{2})}{3!} \delta^3 f_{j+\frac{1}{2}} + \dots ,$$

which is *Bessel's formula*. Bessel's formula is convenient to use for θ near $\frac{1}{2}$, say $\frac{1}{4} \leqslant \theta \leqslant \frac{3}{4}$.

4 EVERETT'S FORMULA

If the formula given in the previous section,

$$f(x_j + \theta h) = f_{j+1} + (\theta - 1)\delta f_{j+\frac{1}{2}} + \frac{\theta(\theta - 1)}{2!} \delta^2 f_{j+1} +$$

$$+ \frac{\theta(\theta - 1)(\theta - 2)}{3!} \delta^3 f_{j+\frac{1}{2}} + \dots ,$$

is subtracted from the Gauss forward formula,

$$f(x_{j+1} + \theta h) = f_{j+1} + \theta \delta f_{j+\frac{3}{2}} + \frac{\theta(\theta - 1)}{2!} \delta^2 f_{j+1} +$$

$$+ \frac{(\theta + 1)\theta(\theta - 1)}{3!} \delta^2 f_{j+\frac{3}{2}} + \dots ,$$

one has

$$f(x_{j+1} + \theta h) - f(x_j + \theta h) = \bar\theta \delta f_{j+\frac{1}{2}} + \frac{(\bar\theta + 1)\bar\theta(\bar\theta - 1)}{3!} \delta^3 f_{j+\frac{1}{2}} + \dots$$

$$+ \theta \delta f_{j+\frac{3}{2}} + \frac{(\theta + 1)\theta(\theta - 1)}{3!} \delta^3 f_{j+\frac{3}{2}} + \dots ,$$

where $\bar\theta \equiv 1 - \theta$. If we define $g_j \equiv f_{j+1} - f_j = \delta f_{j+\frac{1}{2}}$, this last result may be re-expressed as

$$g(x_j + \theta h) = \bar{\theta} g_j + \frac{(\bar{\theta}+1)\bar{\theta}(\bar{\theta}-1)}{3!} \delta^2 g_j + \ldots$$

$$+ \theta g_{j+1} + \frac{(\theta+1)\theta(\theta-1)}{3!} \delta^2 g_{j+1} + \ldots,$$

which is *Everett's formula*.

Everett's formula is occasionally needed: in printed tables it is convenient to give only even differences, and tables of the Everett coefficients are available.

5 FORMULA PATTERNS

Most students find it helpful to remember the pattern traced across a difference table in any application of an interpolation formula.

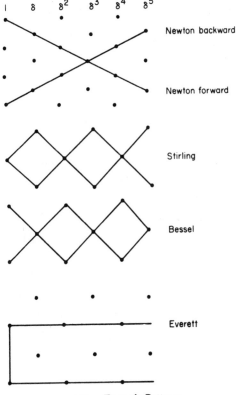

FIGURE 11. Formula Patterns.

The pattern for the Newton formulae should already be familiar, and the patterns for the Stirling, Bessel and Everett formulae may be noted.

6 EXAMPLE

In Section 3 of Step 20, we used Newton's formulae to interpolate for sin 5° and sin 45°. We could use Newton's forward formula to obtain (say)

$$\sin 21° = 0{\cdot}3420 + \tfrac{1}{10}(0{\cdot}1580) + \tfrac{1}{2}\,\tfrac{1}{10}\left(-\tfrac{9}{10}\right)(-0{\cdot}0152)$$

$$\left(-\tfrac{19}{10}\right)(-0{\cdot}0044)$$

$$= 0{\cdot}3420 + 0{\cdot}0158 + 0{\cdot}0006(8) - 0{\cdot}0001(3)$$

$$= 0{\cdot}3584,$$

The same result is obtained using Stirling's formula:

$$\sin 21° = 0{\cdot}3420 + \tfrac{1}{20}\,(0{\cdot}1684 + 0{\cdot}1580) + \tfrac{1}{200}\,(-0{\cdot}0104)$$
$$+ \tfrac{1}{6}\,\tfrac{11}{10}\,\tfrac{1}{10}\left(\tfrac{-9}{10}\right)\tfrac{1}{2}(-0{\cdot}0052 - 0{\cdot}0048)$$
$$= 0{\cdot}3420 - 0{\cdot}0163(2) - 0{\cdot}0000(5) + 0{\cdot}0000(8)$$
$$= 0{\cdot}3584.$$

Again, we could use Newton's forward formula to obtain (say)

$$\sin 25° = 0{\cdot}3420 + \tfrac{1}{2}(0{\cdot}1580) + \tfrac{1}{2}\,\tfrac{1}{2}\left(-\tfrac{1}{2}\right)(-0{\cdot}0152)$$
$$+ \tfrac{1}{6}\,\tfrac{1}{2}\left(-\tfrac{1}{2}\right)\left(-\tfrac{3}{2}\right)(-0{\cdot}0044)$$
$$= 0{\cdot}3420 + 0{\cdot}0790 + 0{\cdot}0019 - 0{\cdot}0002(8)$$
$$= 0{\cdot}4226$$

but there is less computation involved with Bessel's formula:

$$\sin 25° = \tfrac{1}{2}(0{\cdot}3420 + 0{\cdot}5000) + \tfrac{1}{2}\,\tfrac{1}{2}\left(-\tfrac{1}{2}\right)\tfrac{1}{2}(-0{\cdot}0104 - 0{\cdot}0152)$$
$$= 0{\cdot}4210 + 0{\cdot}0016$$
$$= 0{\cdot}4226$$

Finally, note that Everett's formula produces an identical result in this case.

Checkpoint

---

1. When may Stirling's interpolation formula be convenient?
2. When may Bessel's interpolation formula be convenient?
3. Why is Everett's formula used with certain printed tables?

EXERCISES

From a 5D difference table of $f(x) = e^x$ for $x = 0.10\ (0.05)\ 0.40$, estimate

i) $f(0.31)$, using Stirling's formula;

ii) $f(0.31)$, using Everett's formula;

iii) $f(0.315)$, using Bessel's formula;

iv) $f(0.315)$, using Everett's formula.

INTERPOLATION 4
Lagrange interpolation formula

In Steps 19-21 we considered various interpolation formulae that make use of finite differences. Yet another interpolation formula is that attributed to Lagrange, which does not use finite differences at all and has the advantage that it may be used on functions which are not tabulated at equal intervals of the argument.

However, the Lagrange formula is awkward to work with and has a disadvantage in that the degree of the approximating polynomial must be chosen at the outset, so it is mainly of theoretical interest only.

1 PROCEDURE[†]

Suppose that the function $f(x)$ is tabulated at $n+1$ (not necessarily equidistant) points $\{x_0, x_1, ..., x_n\}$ and is to be approximated by a polynomial

$$P_n(x) = a_n x^n + a_{n-1} x^{n-1} + ... + a_1 x + a_0$$

of degree at most n, such that

$$f_j = f(x_j) = P_n(x_j) \text{ for } j = 0, 1, ..., n.$$

Now, for $k = 0, 1, ..., n$.

$$L_k(x) = \frac{(x-x_0)(x-x_1)...(x-x_{k-1})(x-x_{k+1})...(x-x_n)}{(x_k-x_0)(x_k-x_1)...(x_k-x_{k-1})(x_k-x_{k+1})...(x_k-x_n)}$$

is a polynomial of degree n which satisfies

$$L_k(x_j) = 0, j \neq k, j = 0, 1, ..., n,$$

$$L_k(x_k) = 1.$$

Hence

$$P_n(x) = \sum_{k=0}^{n} L_k(x) f_k$$

† This algorithm is suitable for automatic computation. A flow-chart for study and use in programming may be found on page 154.

is a polynomial of degree (at most) n such that

$$P_n(x_j) = f_j, \quad j = 0, 1, \ldots, n;$$

i.e., it is the (unique) collocation polynomial. Note that for $x = x_j$ all terms in the sum vanish except the jth, which is f_j; $L_k(x)$ is called the kth *Lagrange interpolation coefficient*, and the identity $\sum\limits_{k=0}^{n} L_k(x) = 1$ (established by setting $f(x) \equiv 1$) may be used as a check. Note also that with $n = 1$ we recover the linear interpolation formula

$$P_1(x) = \frac{(x-x_1)}{(x_0-x_1)} f_0 + \frac{(x-x_0)}{(x_1-x_0)} f_1 = f_0 + \frac{(x-x_0)}{(x_1-x_0)} (f_1 - f_0)$$

of Step 19.

2   EXAMPLE

Use the Lagrange interpolation formula to find the collocation polynomial $P_3(x)$ through the points $(0, 3)$, $(1, 2)$, $(2, 7)$, $(4, 59)$, and hence find $P_3(3)$.

The Lagrange coefficients are

$$
\begin{aligned}
L_0(x) &= \frac{(x-1)\,(x-2)\,(x-4)}{(0-1)\,(0-2)\,(0-4)} \\
&= -\tfrac{1}{8}(x^3 - 7x^2 + 14x - 8),
\end{aligned}
$$

$$
\begin{aligned}
L_1(x) &= \frac{(x-0)\,(x-2)\,(x-4)}{(1-0)\,(1-2)\,(1-4)} \\
&= \tfrac{1}{3}(x^3 - 6x^2 + 8x),
\end{aligned}
$$

$$
\begin{aligned}
L_2(x) &= \frac{(x-0)\,(x-1)\,(x-4)}{(2-0)\,(2-1)\,(2-4)} \\
&= -\tfrac{1}{4}(x^3 - 5x^2 + 4x),
\end{aligned}
$$

$$
\begin{aligned}
L_3(x) &= \frac{(x-0)\,(x-1)\,(x-2)}{(4-0)\,(4-1)\,(4-2)} \\
&= \tfrac{1}{24}(x^3 - 3x^2 + 2x)
\end{aligned}
$$

(the student should verify that $L_0(x) + L_1(x) + L_2(x) + L_3(x) = 1$). Hence, the required polynomial is

$$P_3(x) = -\tfrac{3}{8}(x^3 - 7x^2 + 14x - 8)$$
$$+ \tfrac{2}{3}(x^3 - 6x^2 + 8x)$$
$$- \tfrac{7}{4}(x^3 - 5x^2 + 4x)$$
$$+ \tfrac{59}{24}(x^3 - 3x^2 + 2x)$$
$$= \tfrac{1}{24}(-9x^3 + 63x^2 - 126x + 72$$
$$+ 16x^3 - 96x^2 + 128x$$
$$- 42x^3 + 210x^2 - 168x$$
$$+ 59x^3 - 177x^2 + 118x)$$
$$= \tfrac{1}{24}(+24x^3 + 0x^2 - 48x + 72)$$
$$= x^3 - 2x + 3.$$

Consequently, $P_3(3) = 27 - 6 + 3 = 24$. However, note that if the explicit form of the collocation polynomial was not required, one would proceed to evaluate $P_3(x)$ for some x directly from the factored forms of $L_k(x)$. Thus, to evaluate $P_3(3)$, one has

$$L_0(3) = \frac{(3-1)(3-2)(3-4)}{(0-1)(0-2)(0-4)} = \frac{1}{4}, \text{ etc.}$$

3 NOTES OF CAUTION

In the case of the Newton or other difference interpolation formulae, the degree of the required approximating polynomial may be determined merely by computing terms until they no longer appear significant. In the Lagrange procedure, the polynomial degree must be chosen at the outset. Also, note that (i) a change of degree involves a completely new computation of all terms; and (ii) for a polynomial of high degree the process involves a large number of multiplications and therefore may be quite slow.

Lagrange interpolation should be used with considerable caution. For example, suppose we use Lagrange interpolation to estimate $\sqrt[3]{20}$ from the points $(0, 0)$, $(1, 1)$, $(8, 2)$, $(27, 3)$, $(64, 4)$ on $f(x) = \sqrt[3]{x}$. We have

$$f(x) \approx \frac{x(x-8)(x-27)(x-64)}{1(-7)(-26)(-63)} \times 1 \;+\; \frac{x(x-1)(x-27)(x-64)}{8(7)(-19)(-56)} \times 2$$
$$+ \frac{x(x-1)(x-8)(x-64)}{27(26)(19)(-37)} \times 3 \;+\; \frac{x(x-1)(x-8)(x-27)}{64(63)(56)(37)} \times 4,$$

so that $f(20) \approx -1.3139$,

which is not very close to the correct value 2·7144! A better result (2·6316) can be obtained by *linear* interpolation between (8, 2) and (27, 3).

The problem is that the Lagrange method gives no indication as to how well $f(x) = \sqrt[3]{x}$ is represented by a quartic. In practice, therefore, the Lagrange interpolation formula is used only rarely (it is much more important in the development of *theory* in Numerical Analysis.)

Checkpoint

1. When is the Lagrange interpolation formula used in practical computation?
2. What distinguishes the Lagrange formula from many other interpolation formulae?
3. Why should the Lagrange formula be used in practice only with caution?

EXERCISE

Given that $f(-2) = 46$, $f(-1) = 4$, $f(1) = 4$, $f(3) = 156$ and $f(4) = 484$, use the Lagrange interpolation formula to compute $f(0)$.

INTERPOLATION 5*
Divided differences and Aitken's method

We noted that the Lagrange interpolation formula is mainly of theoretical interest, for at best it involves very considerable computation in practice, and it can be quite dangerous to use. It is much more efficient to use *divided differences* to interpolate a tabular function with unequally spaced arguments, and at the same time it is relatively safe since the necessary degree of collocation polynomial can be decided. An allied procedure due to Aitken is also commonly adopted in practice.

1 DIVIDED DIFFERENCES

Again, suppose the function $f(x)$ is tabulated at the (not necessarily equidistant) points $\{x_0, x_1, ..., x_n\}$. We define the divided differences between points thus:

first divided difference (say, between x_0 and x_1) by

$$f(x_0, x_1) = \frac{f(x_1)-f(x_0)}{x_1-x_0} = \frac{f_1-f_0}{x_1-x_0} = f(x_1, x_0);$$

second divided difference (say, between x_0, x_1 and x_2) by

$$f(x_0, x_1, x_2) = \frac{f(x_1, x_2)-f(x_0, x_1)}{x_2-x_0};$$

and so on to *nth divided difference* (between x_0, x_1, ..., x_n)

$$f(x_0,x_1,...,x_n) = \frac{f(x_1,x_2,...,x_n)-f(x_0,x_1,...,x_{n-1})}{x_n-x_0}.$$

Example

Construct a divided difference table from the following data:

x	0	1	3	6	10
$f(x)$	1	-6	4	169	921

The difference table is as follows:

x	$f(x)$				
0	1				
		-7			
1	-6		$+4$		
		$+5$		$+1$	
3	4		$+10$		0
		$+55$		$+1$	
6	169		$+19$		
		$+188$			
10	921				

It is notable that the third divided differences are constant. Below, we shall interpolate from the table by using *Newton's divided difference formula*, and determine the corresponding collocation cubic.

2 NEWTON'S DIVIDED DIFFERENCE FORMULA

From the definitions of divided differences, we have

$$f(x) = f(x_0) + (x-x_0)f(x,x_0)$$

$$f(x,x_0) = f(x_0,x_1) + (x-x_1)f(x,x_0,x_1)$$

$$f(x,x_0,x_1) = f(x_0,x_1,x_2) + (x-x_2)f(x,x_0,x_1,x_2)$$

$$\vdots \qquad\qquad \vdots \qquad\qquad \vdots$$

$$f(x,x_0,\ldots,x_{n-1}) = f(x_0,x_1,\ldots,x_n) + (x-x_n)f(x,x_0,x_1,\ldots,x_n).$$

Multiplying the second equation by $(x-x_0)$, the third by $(x-x_0)(x-x_1)$, etc., and adding yields

$$f(x) = f(x_0) + (x-x_0)f(x_0,x_1) + (x-x_0)(x-x_1)f(x_0,x_1,x_2)$$

$$+ \ldots + (x-x_0)(x-x_1)\ldots(x-x_{n-1})f(x_0,x_1,\ldots,x_n) + R$$

where

$$R = (x-x_0)(x-x_1)\ldots(x-x_n)f(x,x_0,x_1\ldots,x_n).$$

Note that the remainder term R is zero at $x_0, x_1, \ldots, x_n$, and we may infer that the other terms on the right hand side constitute the collocation polynomial or, equivalently, the Lagrange polynomial. If the degree of collocation polynomial necessary is not known in advance, it is customary to order the points $x_0, x_1, ..., x_n$ according to increasing distance from x and add terms until R is small enough.

Example

From the tabular function in Section 1 of this Step, find $f(2)$ by Newton's divided difference formula and note the corresponding collocation polynomial. Do the same for $f(4)$.

Since the third difference is constant, we can fit a cubic through the five points. By Newton's divided difference formula, using $x_0 = 0$, $x_1 = 1, x_2 = 3, x_3 = 6$ the cubic is

$$f(x) = f(0) + xf(0, 1) + x(x-1)f(0, 1, 3) + x(x-1)(x-3)f(0, 1, 3, 6)$$
$$= 1 - 7x + 4x(x-1) + 1x(x-1)(x-3),$$

so that

$$f(2) = 1 - 14 + 8 - 2 = -7.$$

The collocation polynomial is obviously

$$1 - 7x + 4x^2 - 4x + x^3 - 4x^2 + 3x = x^3 - 8x + 1.$$

To find $f(4)$, let us identify $x_0 = 1, x_1 = 3, x_2 = 6, x_3 = 10$, so that

$$f(x) = -6 + 5(x-1) + 10(x-1)(x-3) + (x-1)(x-3)(x-6)$$

and

$$f(4) = -6 + 5 \times 3 + 10 \times 3 + 3 \times 1(-2) = +33.$$

As expected, the collocation polynomial is the same cubic — viz., $x^3 - 8x + 1$.

3 AITKEN'S METHOD

In practice, a procedure due to Aitken is often adopted, in which successively better interpolation polynomials (corresponding to successively higher order truncation of Newton's divided difference formula) are determined systematically. Thus, one has

$$f(x) \approx f_0 + (x-x_0)\frac{f_1-f_0}{x_1-x_0}$$
$$= \frac{f_0(x_1-x)-f_1(x_0-x)}{x_1-x_0} \equiv I_{0,1}(x),$$

and obviously

$$f_0 = I_{0,1}(x_0), \qquad f_1 = I_{0,1}(x_1).$$

Next, one has

$$f(x) \approx f_0 + (x - x_0)f(x_0,x_1) + (x - x_0)(x - x_1)\frac{f(x_0,x_2) - f(x_0,x_1)}{x_2 - x_1}$$

$$= \frac{I_{0,1}(x)(x_2 - x) - I_{0,2}(x)(x_1 - x)}{x_2 - x_1} \equiv I_{0,1,2}(x),$$

noting that

$$I_{0,2}(x) = f_0 + (x - x_0)f(x_0,x_2),$$

and so on. In passing, one may note that

$$f_0 = I_{0,1,2}(x_0), \qquad f_1 = I_{0,1,2}(x_1), \qquad f_2 = I_{0,1,2}(x_2).$$

At first sight, the procedure may look complicated, but it is systematic, and therefore computationally straight forward: it may be represented by the scheme

x_0	f_0					$x_0 - x$
x_1	f_1	$I_{0,1}(x)$				$x_1 - x$
x_2	f_2	$I_{0,2}(x)$	$I_{0,1,2}(x)$			$x_2 - x$
x_3	f_3	$I_{0,3}(x)$	$I_{0,1,3}(x)$	$I_{0,1,2,3}(x)$		$x_3 - x$
....	....	....	....	....		....

One major advantage is that the computor may gauge the accuracy achieved by comparing successive steps. (This of course corresponds to gauging the appropriate truncation of the Newton divided difference formula.) As in the case of Newton's divided difference formula, usually the points $x_0, x_1, x_2 \ldots$ are ordered such that $x_0 - x, x_1 - x, x_2 - x, \ldots$ form a sequence with increasing magnitude.

Finally, we remark that although the derivation of Aitken's method emphasizes its relationship with the Newton formula, it is notable that Aitken's method ultimately does not involve divided differences at all!

Example

From the tabular function in Section 1 of this Step, find $f(2)$ by Aitken's method.

We have $x = 2$, so that we choose $x_0 = 1$, $x_1 = 3$, $x_2 = 0$, $x_3 = 6$, $x_4 = 10$:

	x_k	f_k					$x_k - x$
$k = 0$	1	-6					-1
$k = 1$	3	4	-1				$+1$
$k = 2$	0	1	-13	-5			-2
$k = 3$	6	169	29	-11	-7		$+4$
$k = 4$	10	921	97	-15	-7	-7	$+8$

The computation proceeds from the left, row by row, with an appropriately divided 'cross multiplication' of the respective entries with those in the $(x_k - x)$ column on the right: thus,

$$I_{0,1} = \frac{(-6)(+1) - (+4)(-1)}{3 - 1} = -1$$

$$I_{0,2} = \frac{(-6)(-2) - (+1)(-1)}{0 - 1} = -13$$

$$I_{0,1,2} = \frac{(-1)(-2) - (-13)(+1)}{0 - 3} = -5$$

etc.

The entry -7 (ringed) appears twice successively along the diagonal, so that one may conclude that $f(2) = -7$.

Checkpoint

1. What major practical advantage has Newton's divided difference interpolation formula over Lagrange's formula?
2. How are the tabular points usually ordered for interpolation by either Newton's divided difference formula or Aitken's method?
3. Are divided differences actually used in interpolation by Aitken's method?

EXERCISES

1. Use Newton's divided difference formula to show that an interpolation for $\sqrt[3]{20}$ from the points $(0, 0)$, $(1, 1)$, $(8, 2)$, $(27, 3)$, $(64, 4)$, on $f(x) = \sqrt[3]{x}$ is quite invalid.

2. Given that $f(-2) = 46$, $f(-1) = 4$, $f(1) = 4$, $f(3) = 156$ and $f(4) = 484$, compute $f(0)$ from (i) Newton's divided difference formula; and (ii) Aitken's method.

   Comment on the validity of this interpolation.

3. Given that $f(0) = 2 \cdot 3913$, $f(1) = 2 \cdot 3919$, $f(3) = 2 \cdot 3938$ and $f(4) = 2 \cdot 3951$, use Aitken's method to estimate $f(2)$.

INTERPOLATION 6*
Inverse interpolation

Rather than the value of a function $f(x)$ for a certain x, one might seek the value of x corresponding to a given value of $f(x)$; this is called *inverse interpolation*. For example, perhaps the student may have contemplated the possibility of obtaining roots of $f(x) = 0$ by inverse interpolation.

1 LINEAR INVERSE INTERPOLATION

An obvious elementary procedure is to tabulate the function in the neighbourhood of the given value at an interval so small that *linear inverse interpolation* may be used. Reference to Step 19 provides

$$x = x_j + \theta(x_{j+1} - x_j),$$

where

$$\theta \approx \frac{f(x) - f_j}{f_{j+1} - f_j}$$

in the linear approximation. (Note that if $f(x) = 0$, we recover the method of false position — see Step 8).

For example, from a 4D table of $f(x) = e^{-x}$ one has $f(0\cdot91) = 0\cdot4025$, $f(0\cdot92) = 0\cdot3985$ so that $f(x) = 0\cdot4$ corresponds to

$$x \approx 0\cdot91 + \frac{0\cdot4 - 0\cdot4025}{0\cdot3985 - 0\cdot4025} \times (0\cdot92 - 0\cdot91)$$

$$= 0\cdot91 + 0\cdot00625 = 0\cdot91625$$

An immediate check is to use (direct) interpolation to recover $f(x) = 0\cdot4$. Thus,

$$f(0\cdot91625) \approx 0\cdot4025 + \frac{0\cdot91625 - 0\cdot91}{0\cdot92 - 0\cdot91} \times (0\cdot3985 - 0\cdot4025)$$

$$= 0\cdot4000 \, .$$

2 ITERATIVE INVERSE INTERPOLATION

As no doubt the student may appreciate, it may be preferable to adopt
(at least approximately) a collocation polynomial of degree greater
than one, rather than seek to tabulate at a small enough interval to
permit linear inverse interpolation. The degree of the approximating
polynomial may be decided implicitly by an iterative (successive
approximation) method.

For example, Newton's forward difference formula may be rearranged as

$$\theta = \{f(x) - f_j - \tfrac{1}{2}\theta(\theta-1)\Delta^2 f_j + \ldots\}/\Delta f_j.$$

Since terms involving second and higher differences may be expected
to decrease fairly quickly, we have successive approximations $\{\theta_s\}$ to
θ given by

$$\theta_1 = \{f(x) - f_j\}/\Delta f_j,$$
$$\theta_2 = \{f(x) - f_j - \tfrac{1}{2}\theta_1(\theta_1-1)\Delta^2 f_j\}/\Delta f_j,$$

etc.

Similar iterative procedures may be based on other difference formulae;
for example, Everett's formula produces

$$\theta_1 = \{f(x) - f_j\}/\delta f_{j+\frac{1}{2}},$$
$$\theta_2 = \{f(x) - f_j + \tfrac{1}{6}\theta_1(\theta_1-1)[(\theta_1-2)\delta^2 f_j - (\theta_1+1)\delta^2 f_{j+1}]\}/\delta f_{j+\frac{1}{2}},$$

etc.

To illustrate, consider the table of $f(x) = \sin x$ for $x = 0° (10°) 50°$
given in Step 20, and suppose we seek x for $f(x) = 0·2$. Since some
electronic desk calculators do not have inverse trigonometric function
keys, this simple example is of practical interest. Clearly, $10° < x < 20°$.
From Newton's formula,

$$\theta_1 = (0·2 - 0·1736)/0·1684 = 0·0264/0·1684 = 0·1568 \approx 0·16$$

$$\theta_2 = \{0·0264 - \tfrac{1}{2}(0·16)(-0·84)(-0·0104)\}/0·1684$$

$$= \{0·0264 - 0·0007\}/0·1684 = 0·1526 \approx 0·153$$

$$\theta_3 = \{0.0264 - \tfrac{1}{2}(0.153)(-0.847)(-0.0104)$$
$$\qquad - \tfrac{1}{6}(0.153)(-0.847)(-1.847)(-0.0048)\}/0.1684$$
$$= \{0.0264 - 0.0007 + 0.0002\}/0.1684$$
$$= 0.1538.$$

(Note that it is unneccessary to carry many figures in the first estimates of θ.) From Everett's formula,

$$\theta_1 = 0.16$$
$$\theta_2 = \{0.0264 + \tfrac{1}{6}(0.16)(-0.84)[(-1.84)(-0.0052)$$
$$\qquad -(1.16)(-0.0104)]\}/0.1684$$
$$= \{0.0264 - (0.0224)(0.0216)\}/0.1684$$
$$= \{0.0264 - 0.0005\}/0.1684$$
$$= 0.1538.$$

Consequently, $\theta = 0.1538 = \dfrac{x - 10}{10}$ yields $x = 11.538°$.

Checking, either by the usual method of direct interpolation or in this case directly, yields $\sin 11.538° = 0.2000$.

3   DIVIDED DIFFERENCES

Since divided differences are suitable for interpolation with unequally spaced tabular values, they may be used for inverse interpolation. Let us again consider the function $f(x) = \sin x$ for $x = 0°$ $(10°)$ $50°$, and determine x for $f(x) = 0.2$. Ordering with increasing distance from $f(x) = 0.2$, one has the divided difference table (entries multiplied by 100):

$f(x)$	x					
0.1736	10					
		5938				
0.3420	20		518			
		5848		1360		
0.0000	0		962		1338	
		6000		1988		3486
0.5000	30		1560		3403	
		7003		3431		
0.6428	40		4188			
		8117				
0.7660	50					

Consequently,

$$x = 10 + (0.2 - 0.1736)59.38$$

$$+ (0.2 - 0.1736)(0.2 - 0.3420)5.18$$

$$+ (0.2 - 0.1736)(0.2 - 0.3420)(0.2 - 0)13.60$$

$$+ (0.2 - 0.1736)(0.2 - 0.3420)(0.2 - 0)(0.2 - 0.5)13.38$$

$$+ (0.2 - 0.1736)(0.2 - 0.3420)(0.2 - 0)(0.2 - 0.5)$$
$$\times (0.2 - 0.6428)34.86$$

$$= 10 + 1.5676 - 0.0194 - 0.0102 + 0.0030 - 0.0035$$

$$= 11.5375.$$

Alternatively, the Aitken scheme could be used. With either alternative, however, it is noticeable that any advantage in accuracy compared with iterative inverse interpolation may not justify the additional computational demand.

Checkpoint

1. Why may linear inverse interpolation be either tedious or impractical?
2. What is the usual method for checking inverse interpolation?
3. What potential advantage has inverse interpolation using either divided differences or the Aitken scheme, compared with the iterative method? What is a likely disadvantage?

EXERCISES

1. Use linear inverse interpolation to find the root of $x + \cos x = 0$.
2. Solve $3xe^x = 1$ to 3D.
3. The corrected table for the cubic $f(x)$ given in Exercise 1 of Step 18 is

x	$f(x)$	x	$f(x)$	x	$f(x)$	x	$f(x)$
2	3·0671	5	13·3184	8	24·2573	11	35·9000
3	6·4088	6	16·8875	9	28·0592	12	39·9401
4	9·8257	7	20·5336	10	31·9399	13	44·0608

Without knowledge of the explicit form of $f(x)$, find x for $f(x) = 10, 20$ and 40 respectively. Check your answers by (direct) interpolation. Finally, obtain the equation of the cubic, and use it to check your answers again.

CURVE FITTING

Scientists and social scientists often wish to fit a smooth curve to some experimental data. Given $(n + 1)$ points an obvious approach is to use the interpolating polynomial of degree n, but when n is large this is usually unsatisfactory. Better results can be obtained by using *piecewise* polynomials, i.e. fitting lower degree polynomials through subsets of the data points. The use of *spline* functions (which usually provide a particularly smooth fit) has become widespread (see for example Conte and de Boor).

A rather different and frequently more effective type of curve fitting is the *least squares fit* in which, rather than try to fit the points *exactly*, we find a polynomial of low degree (often first or second) which fits the points *closely* (after all, the points themselves will not be exact, being subject to experimental error).

1 THE PROBLEM ILLUSTRATED

Suppose we are studying experimentally the relationship between two variables x and y—for example, quantity (x) of drug injected and observed reaction time (y), measured in a sample of rats. By carrying out the appropriate experiment six times (say), we obtain six pairs of values (x_i, y_i); these can be plotted on a diagram such as Figure 12(a).

We may believe that the relationship between the variables can be satisfactorily described by a functional relation $y = f(x)$, but that the y-values obtained experimentally are subject to *errors* (or *noise*). The mathematical model of this situation is as follows:

$$f(x_i) = y_i + \varepsilon_i \quad (i = 1, ..., n, \text{ when there are } n \text{ observed points}).$$

Here $f(x_i)$ is the functional value of y corresponding to the value x_i used in the experiment, and ε_i is the experimental error involved in the measurement of the y-variable at the point. Thus the error in y at the observed point is $\varepsilon_i = f(x_i) - y_i$.

The problem of curve-fitting is to use the information of the sample data points to determine a suitable curve (i.e. find a suitable function $f(x)$) so that the equation $y = f(x)$ can be used as a description of the

(x,y) relationship; the hope is that predictions made from this equation will not be too much in error.

How is the function $f(x)$ to be chosen? There is an unlimited number of functions from which to choose, and Figure 12(b) shows four possibilities. The polygon A passes through all six points; intuitively, however, we might prefer to fit a straight line such as B, or an exponential curve such as C. The curve D is clearly not a good candidate for our model.

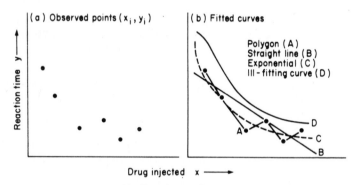

FIGURE 12. Reaction to drugs in rats

2 GENERAL APPROACH TO THE PROBLEM

To begin to answer the question of which function to choose, we state one possible approach.

Given a set of values $(x_1, y_1), (x_2, y_2) \ldots, (x_n, y_n)$, we shall pick a function which we can specify completely except for the values of a set of k parameters $c_1, c_2, \ldots, c_k$; we denote this function $y = f(x; c_1, c_2, \ldots, c_k)$.

We shall choose values for the parameters which will make the errors at the measured points (x_i, y_i) as small as possible. We shall suggest three ways by which the phrase 'as small as possible' can be given specific meaning.

Examples of functions to use

i) $y = c_1 + c_2 x + c_3 x^2 + \ldots + c_k x^{k-1}$ (polynomial)

ii) $y = c_1 \sin \omega x + c_2 \sin 2\omega x + \ldots + c_k \sin k\omega x$
       (combination of sine functions)

iii) $y = c_1 \cos \omega x + c_2 \cos 2\omega x + \ldots + c_k \cos k\omega x$
       (combination of cosine functions)

Examples (i), (ii) and (iii) are themselves examples of what may be termed the *general linear form*:

iv) $y = c_1\phi_1(x) + c_2\phi_2(x) + \ldots + c_k\phi_k(x),$

where the functions $\phi_1(x)$, $\phi_2(x)$, ..., $\phi_k(x)$ constitute a pre-selected set of functions. In (i) the set of functions is $\{1, x, x^2, \ldots, x^{k-1}\}$; in (ii) it is $\{\sin \omega x, \sin 2\omega x, \ldots, \sin k\omega x\}$, with ω being a constant chosen to agree with a periodicity in the data; in (iii) the set is $\{\cos \omega x, \cos 2\omega x, \ldots, \cos k\omega x\}$. Other functions commonly used in curve-fitting are exponential functions, Bessel functions, Legendre polynomials, and Chebyshev polynomials (see for example Conte and de Boor).

3   ERRORS 'AS SMALL AS POSSIBLE'

We will present criteria which make precise the concept of choosing a function to make measurement errors as small as possible. We suppose that the curve to be fitted can be expressed in general linear form, with a known set of functions $\{\phi_1(x), \phi_2(x), \ldots, \phi_k(x)\}$.

The errors $\varepsilon_i = y - y_i$ at the n data points are as follows:

$$\varepsilon_1 = c_1\phi_1(x_1) + c_2\phi_2(x_1) + \ldots + c_k\phi_k(x_1) - y_1$$

$$\varepsilon_2 = c_1\phi_1(x_2) + c_2\phi_2(x_2) + \ldots + c_k\phi_k(x_2) - y_2$$

$$\varepsilon_n = c_1\phi_1(x_n) + c_2\phi_2(x_n) + \ldots + c_k\phi_k(x_n) - y_n$$

If the number of data-points is less than or equal to the number of parameters (i.e. $n \leqslant k$), it is possible to find values for $\{c_1, c_2, \ldots, c_k\}$ which make all the errors ε_i zero. If $n < k$ there is an infinite number of solutions for $\{c_i\}$ which make all the errors zero (and therefore an infinite number of curves with the given form pass through all the experimental points); in this case the problem is not fully determined — more information is needed to choose an appropriate curve.

If $n > k$, which in practice is usually the case, the errors cannot all be made zero by a choice of the $\{c_i\}$. Three possible procedures are as follows:

i)   choose a set $\{c_i\}$ which minimizes the total absolute error;

i.e. minimize the sum $\displaystyle\sum_{i=1}^{n} |\varepsilon_i|$;

ii)   choose a set $\{c_i\}$ which minimizes the maximum absolute error;

i.e. minimize $\underset{i=1,\ldots,n}{\text{Max}} \{|\varepsilon_i|\}$;

iii) choose a set $\{c_i\}$ which minimizes the sum of squares of errors;

i.e. minimize $S' = \displaystyle\sum_{i=1}^{n} \varepsilon_i^2$

Procedures (i) and (ii) are generally difficult to apply. Procedure (iii) leads to a linear system of equations to solve for the set $\{c_i\}$; it is called the *principle of least squares*, and is the one customarily used.

4   THE LEAST SQUARES METHOD

In order to apply the principle of least squares it is necessary to use partial differentiation, a calculus technique which may not be known to readers of this text. For that reason a general description of the method will not be given, but an outline will be attempted, together with examples of its use.

a)   *The normal equations*

The sum of squared errors to be minimized is

$$S = \sum_{i=1}^{n} \varepsilon_i^{2} = \sum_{i=1}^{n} [f(x_i;c_1,c_2,...,c_k) - y_i]^2$$

The n values of (x_i, y_i) are the known measurements taken from n experiments. When they are inserted in the right hand side, S becomes an expression involving only k unknowns, namely $c_1, c_2, ..., c_k$. In other words S may be regarded as a function of the c_i; i.e. $S \equiv S(c_1, c_2, ..., c_k)$. The problem is to choose that set of values for $\{c_i\}$ which makes S a minimum.

A theorem in calculus tells us that (under certain conditions which are usually satisfied in practice) the minimum of S occurs at the point where all the *partial derivatives* $\dfrac{\partial S}{\partial c_1}, \dfrac{\partial S}{\partial c_2}, ..., \dfrac{\partial S}{\partial c_k}$ vanish. (The partial derivative $\dfrac{\partial S}{\partial c_1}$ (for example) is the same as the differential coefficient $\dfrac{dS}{dc_1}$ with all the other c_i held constant;

e.g., if $S = 3c_1 + 5c_2$, $\dfrac{\partial S}{\partial c_1} = 3$ and $\dfrac{\partial S}{\partial c_2} = 5$.)

Thus we have to solve the following system of k equations:

$$\left.\begin{array}{l} \dfrac{\partial S}{\partial c_1} = 0 \\[3mm] \dfrac{\partial S}{\partial c_2} = 0 \\[2mm] \vdots \\[2mm] \dfrac{\partial S}{\partial c_k} = 0 \end{array}\right\}$$

This system is a set of equations linear in the variables $c_1, c_2, \ldots, c_k$ and they are called the *normal equations* for the least squares approximation. They may be solved by the numerical methods presented in Steps 11 and 13. Their simultaneous solution provides the required set $\{c_i\}$ which minimizes S.

b) *Examples of the least squares method*

The following points were obtained in an experiment:

x	1	2	3	4	5	6
y	1	3	4	3	4	2

Plot the points on a diagram, and use the method of least squares to fit (i) a straight line, and (ii) a parabola through them.

The plotted points are shown in Figure 13(a).

i) *To fit a straight line*, we have to find a function $y = c_1 + c_2 x$ (i.e. a first degree polynomial) which minimizes

$$S = \sum_{i=1}^{6} \varepsilon_i^2 = \sum_{i=1}^{6} (y_i - c_1 - c_2 x_i)^2.$$

Differentiating first with respect to c_1 (keeping c_2 constant) and then with respect to c_2 (keeping c_1 constant), and setting the results equal to zero, gives the normal equations:

$$\left.\begin{array}{l} \dfrac{\partial S}{\partial c_1} \equiv -2 \sum_{i=1}^{6} (y_i - c_1 - c_2 x_i) = 0 \\[5mm] \dfrac{\partial S}{\partial c_2} \equiv -2 \sum_{i=1}^{6} x_i(y_i - c_1 - c_2 x_i) = 0 \end{array}\right\}$$

We may divide both equations by -2, take the sigma operations through the brackets, and rearrange, to obtain:

$$\left. \begin{array}{l} \Sigma y_i = 6c_1 + (\Sigma x_i)c_2 \\ \Sigma x_i y_i = (\Sigma x_i)c_1 + (\Sigma x_i^2)c_2 \end{array} \right\}$$

It is seen that to proceed to a solution we have to evaluate the four sums Σx_i, Σy_i, Σx_i^2, and $\Sigma x_i y_i$, and insert them in the last equations. We can arrange the work in a table thus (the last three columns are for fitting the parabola).

	x	y	x^2	xy	$x^2 y$	x^3	x^4
	1	1	1	1	1	1	1
	2	3	4	6	12	8	16
	3	4	9	12	36	27	81
	4	3	16	12	48	64	256
	5	4	25	20	100	125	625
	6	2	36	12	72	216	1296
Required sums	21	17	91	63	269	441	2275

The normal equations for fitting the straight line are, hence:

$$\left. \begin{array}{l} 17 = 6c_1 + 21c_2 \\ 63 = 21c_1 + 91c_2 \end{array} \right\}$$

The solution to 2D is

$$\left. \begin{array}{l} c_1 = 2 \cdot 13 \\ c_2 = 0 \cdot 20 \end{array} \right\}$$

Consequently, the required line is

$$y = 2 \cdot 13 + 0 \cdot 2x$$

and is plotted in Figure 13(b).

ii) *To fit a parabola* we have to find the second degree polynomial

$$y = c_1 + c_2 x + c_3 x^2$$

which minimizes

$$S = \Sigma \varepsilon_i^2 = \Sigma (y_i - c_1 - c_2 x_i - c_3 x_i^2)^2.$$

Taking partial derivatives and proceeding as above we obtain the normal equations

$$\left.\begin{array}{l} \Sigma y_i = 6c_1 + (\Sigma x_i)c_2 + (\Sigma x_i{}^2)c_3 \\ \Sigma x_i y_i = (\Sigma x_i)c_1 + (\Sigma x_i{}^2)c_2 + (\Sigma x_i{}^3)c_3 \\ \Sigma x_i{}^2 y_i = (\Sigma x_i{}^2)c_1 + (\Sigma x_i{}^3)c_2 + (\Sigma x_i{}^4)c_3 \end{array}\right\}$$

Inserting the values for the sums (see the above table) we obtain the system of linear equations:

$$\left.\begin{array}{l} 17 = 6c_1 + 21c_2 + 91c_3 \\ 63 = 21c_1 + 91c_2 + 441c_3 \\ 269 = 91c_1 + 441c_2 + 2275c_3 \end{array}\right\}$$

The solution, to 3D is

$$\left.\begin{array}{l} c_1 = -1{\cdot}200 \\ c_2 = 2{\cdot}700 \\ c_3 = -0{\cdot}357 \end{array}\right\}$$

The required parabola is therefore (retaining 2D):

$$y = -1{\cdot}20 + 2{\cdot}70x - 0{\cdot}36x^2 \ ;$$

it is also plotted in Figure 13(b). It is clear that the parabola is a better fit than the straight line!

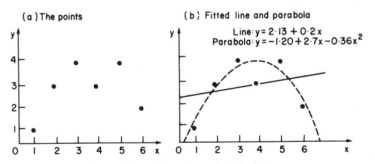

FIGURE 13. Fitting a line and parabola by least squares

Checkpoint

1. What is meant by 'error' at a point?
2. Give three criteria which may be applied to choose the set $\{c_i\}$.
3. How are the normal equations obtained?

EXERCISES

1. For the example of 4(b) above, compute the value of S, the sum of squares of errors of points from (i) the fitted line, and (ii) the fitted parabola. Plot the points on graph paper, and fit a straight line 'by eye' (i.e. use a ruler to draw a line, guessing its best position); determine the value of S for this line and compare with the value for the least squares line.

2. Fit a straight line by the least squares method to each of the following sets of data:
   i) toughness (x) and percentage of nickel (y) in eight specimens of alloy steel.

toughness (x)	36	41	42	43	44	45	47	50
% nickel (y)	2·5	2·7	2·8	2·9	3·0	3·2	3·3	3·5

   ii) aptitude·test mark (x) given to six trainee salesmen, and their first-year sales (y) in hundreds of dollars.

aptitude test (x)	25	29	33	36	42	54
first-year sales (y)	42	45	50	48	73	90

   For both sets, plot the points and draw the least squares line on a graph. Use the lines to predict (i) the % nickel of a specimen of steel whose toughness is 38, and (ii) the likely first-year sales of a trainee salesman who obtains a mark of 48 on his aptitude test.

3. Obtain the normal equations for fitting a third-degree polynomial
   $y = c_1 + c_2x + c_3x^2 + c_4x^3$ to a set of n points. Show that they can be written in matrix form thus:

$$
\begin{bmatrix} \Sigma y \\ \Sigma xy \\ \Sigma x^2 y \\ \Sigma x^3 y \end{bmatrix} = \begin{bmatrix} n & \Sigma x & \Sigma x^2 & \Sigma x^3 \\ \Sigma x & \Sigma x^2 & \Sigma x^3 & \Sigma x^4 \\ \Sigma x^2 & \Sigma x^3 & \Sigma x^4 & \Sigma x^5 \\ \Sigma x^3 & \Sigma x^4 & \Sigma x^5 & \Sigma x^6 \end{bmatrix} \begin{bmatrix} c_1 \\ c_2 \\ c_3 \\ c_4 \end{bmatrix}
$$

   Deduce the matrix form of the normal equations for fitting a fourth-degree polynomial.

4. Fit a parabola by the least squares method to the points (0, 0), (1, 1), (2, 3), (3, 3) and (4, 2). Find the value of S for this fit.

5. Find the normal equations for fitting, by the least squares method, an equation of form $y = c_1 + c_2 \sin x$ to the set of points (0, 0), $(\pi/6, 1)$, $(\pi/2, 3)$ and $(5\pi/6, 2)$. Solve them for c_1 and c_2.

NUMERICAL DIFFERENTIATION

In Analysis, we are usually able to obtain the derivative of a function by the methods of elementary calculus. If the function is very complicated or known only as a table, however, it may be necessary to resort to *numerical differentiation*.

1 PROCEDURE

Formulae for numerical differentiation may easily be obtained by differentiating the interpolation polynomial. The essential idea is that the derivatives $f'(x), f''(x), \ldots$ of a function $f(x)$ are represented by the derivatives $P_n'(x), P_n''(x), \ldots$ of the collocation polynomial $P_n(x)$.

For example, differentiating the Newton forward difference formula

$$f(x) = f(x_j + \theta h) \approx \left[1 + \theta\Delta + \tfrac{1}{2}\theta(\theta-1)\Delta^2 + \frac{\theta(\theta-1)(\theta-2)}{3!}\Delta^3 + \ldots\right]f_j$$

with respect to x gives formally (since $x = x_j + \theta h, \dfrac{\mathrm{d}f}{\mathrm{d}x} = \dfrac{\mathrm{d}f}{\mathrm{d}\theta} \times \dfrac{\mathrm{d}\theta}{\mathrm{d}x}$, etc.)

$$f'(x) = \frac{1}{h}\frac{\mathrm{d}f}{\mathrm{d}\theta} \approx \frac{1}{h}\left[\Delta + (\theta-\tfrac{1}{2})\Delta^2 + \frac{3\theta^2 - 6\theta + 2}{6}\Delta^3 + \ldots\right]f_j$$

$$f''(x) = \frac{1}{h^2}\frac{\mathrm{d}^2f}{\mathrm{d}\theta^2} \approx \frac{1}{h^2}\left[\Delta^2 + (\theta-1)\Delta^3 + \ldots\right]f_j, \qquad \text{etc.}$$

In particular, if we set $\theta = 0$ we have formulae for derivatives at the tabular points $\{x_j\}$:

$$f'_j \approx \frac{1}{h}\left[\Delta - \tfrac{1}{2}\Delta^2 + \tfrac{1}{3}\Delta^3 - \ldots\right]f_j$$

$$f''_j \approx \frac{1}{h^2}\left[\Delta^2 - \Delta^3 + \tfrac{11}{12}\Delta^4 - \ldots\right]f_j, \qquad \text{etc.}$$

(In Step 20 we noted that E is analogous to e^{hD}; we can 'deduce' that $hD = \log_e E = \log_e(1+\Delta) = \Delta - \tfrac{1}{2}\Delta^2 + \tfrac{1}{3}\Delta^3 - \ldots$!)

If we set $\theta = \frac{1}{2}$, we have a relatively accurate formula at half-way points (without second differences)

$$f'_{j+\frac{1}{2}} \approx \frac{1}{h} \left[\Delta - \tfrac{1}{24} \Delta^3 + \ldots \right] f_j ;$$

if we set $\theta = 1$ in the formula for the second derivative, we have the result (without third differences)

$$f''_{j+1} \approx \frac{1}{h^2} \left[\Delta^2 - \tfrac{1}{12} \Delta^4 + \ldots \right] f_j,$$

a formula for the second derivative at the next point.

Note that if only one term is retained, the well-known formulae

$$f'(x_j) \approx \frac{f(x_j + h) - f(x_j)}{h} ;$$

$$f''(x_j) \approx \frac{f(x_j + 2h) - 2f(x_j + h) + f(x_j)}{h^2} \qquad (\theta = 0) ;$$

$$f'(x_j + \tfrac{1}{2}h) \approx \frac{f(x_j + h) - f(x_j)}{h} \qquad (\theta = \tfrac{1}{2}) ;$$

$$f''(x_j + h) \approx \frac{f(x_j + 2h) - 2f(x_j + h) + f(x_j)}{h^2} \qquad (\theta = 1) ;$$

etc. are recovered.

2   ERROR IN NUMERICAL DIFFERENTIATION

It must be recognized that numerical differentiation is subject to considerable error; the basic difficulty is that while $(f(x) - P_n(x))$ may be small, the differences $(f'(x) - P_n'(x))$ and $(f''(x - P_n''(x))$ etc. may be very large. In geometrical terms, although two curves may be close together, they may differ considerably in slope, variation in slope, etc. (see Figure 14).

It should also be noted that the formulae all involve dividing a combination of differences (which are prone to cancellation errors, especially if h is small), by a positive power of h. Consequently if we want to keep *round-off* errors down, we should use a *large* value of h. On the other hand, it can be shown (see Exercise 3 below) that the *truncation* error is approximately proportional to h^p, where p is a positive integer, so that h must be sufficiently *small* for the truncation error to be tolerable. We are in a 'cleft stick' and must compromise with some optimum choice of h.

In brief, large errors may occur in numerical differentiation based on

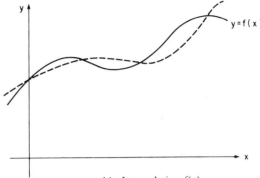

FIGURE 14. Interpolating $f(x)$

direct polynomial approximation, so that an error check is always advisable. There are alternative methods based on polynomials which use more sophisticated procedures such as least-squares or mini-max, and other alternatives involving other base functions (e.g. trigonometric functions). However, the best policy is probably to use numerical differentiation only when it cannot be avoided!

3   EXAMPLE

Estimate $f'(0\cdot1)$ and $f''(0\cdot1)$ for $f(x) = e^x$ using the data in Step 17 (p. 79).

If we use the formulae from p. 125 (with $\theta = 0$) we obtain (ignoring fourth and higher differences):

$$f'(0\cdot1) \approx \frac{1}{0\cdot05} \left[0\cdot05666 - \tfrac{1}{2}(0\cdot00291) + \tfrac{1}{3}(0\cdot00015)\right]$$
$$= 20\,(0\cdot05666 - 0\cdot00145(5) + 0\cdot00005)$$
$$= 1\cdot1051$$

$$f''(0\cdot1) \approx 400\,(0\cdot00291 - 0\cdot0015)$$
$$= 1\cdot104$$

Since $f''(0\cdot1) = f'(0\cdot1) = f(0\cdot1) = 1\cdot10517$, it is obvious that the second result is much less accurate (because of roundoff errors).

Checkpoint

1.   How are formulae for the derivatives of a function obtained from interpolation formulae?
2.   Why is the accuracy of the usual numerical differentiation process *not* necessarily increased if the argument interval is reduced?
3.   When should numerical differentiation be used?

EXERCISES

1. Derive formulae involving backward differences for the first and second derivatives of a function.

2. The function $f(x) = \sqrt{x}$ is tabulated for $x = 1\cdot00$ $(0\cdot05)$ $1\cdot30$ to 5 decimal places:

x	$f(x)$
$1\cdot00$	$1\cdot00000$
$1\cdot05$	$1\cdot02470$
$1\cdot10$	$1\cdot04881$
$1\cdot15$	$1\cdot07238$
$1\cdot20$	$1\cdot09545$
$1\cdot25$	$1\cdot11803$
$1\cdot30$	$1\cdot14018$

   i)   Use Newton's forward difference formula to estimate $f'(1), f''(1)$.

   ii) Use Newton's backward difference formula to estimate $f'(1\cdot30), f''(1\cdot30)$.

3. Use the Taylor series to find the truncation errors in the formulae

   i) $f'(x_j) \approx (f(x_j + h) - f(x_j))/h$;

   ii) $f'(x_j + \tfrac{1}{2}h) \approx (f(x_j + h) - f(x_j)/h$;

   iii) $f''(x_j) \approx (f(x_j + 2h) - 2f(x_j + h) + f(x_j))/h^2$;

   iv) $f''(x_j + h) \approx (f(x_j + 2h) - 2f(x_j + h) + f(x_j))/h^2$.

NUMERICAL INTEGRATION 1
The trapezoidal rule

It is often either difficult or impossible to evaluate definite integrals of the form

$$\int_a^b f(x)\mathrm{d}x$$

by analytical methods, so *numerical integration* or *quadrature* is used.

It is well known that the definite integral may be interpreted as the area under the curve $y = f(x)$ for $a \leqslant x \leqslant b$, and may be evaluated by subdivision of the interval and summation of the component areas. This additive property of the definite integral permits evaluation in a piecewise sense. We may adopt the polynomial approximation $f(x) \approx P_n(x)$ for any sub-interval $x_0 \leqslant x \leqslant x_n$ of the interval $a \leqslant x \leqslant b$ in order to evaluate

$$\int_{x_0}^{x_n} P_n(x)\mathrm{d}x$$

to good approximation, by choosing n to ensure that the error $(f(x) - P_n(x))$ in each tabular sub-interval $x_j \leqslant x \leqslant x_{j+1}$ is sufficiently small. It is notable that (for $n > 1$) the error is often alternately positive and negative in successive sub-intervals, and considerable cancellation of error occurs; in contrast with numerical differentiation quadrature is inherently accurate! It is usually sufficient to use a rather low degree polynomial approximation over any sub-interval $x_0 \leqslant x \leqslant x_n$.

1 THE TRAPEZOIDAL RULE[†]

Perhaps the most straightforward quadrature is to divide the interval $a \leqslant x \leqslant b$ into N equal strips of width h by the points

$$x_j = a + jh, \qquad j = 0, 1, 2, ..., N,$$

such that $b = a + Nh$. Then we can use the additive property

---

† This algorithm is suitable for automatic computation. A flow-chart for study and use in programming may be found on page 155.

$$\int_a^b f(x)\mathrm{d}x = \int_{x_0}^{x_1} f(x)\mathrm{d}x + \int_{x_1}^{x_2} f(x)\mathrm{d}x + \dots + \int_{x_{N-1}}^{x_N} f(x)\mathrm{d}x$$

and the linear approximations (involving $x = x_j + \theta h$)

$$\int_{x_j}^{x_{j+1}} f(x)\mathrm{d}x = \int_{x_j}^{x_j+h} f(x_j + \theta h)\mathrm{d}x \qquad = h\int_0^1 f(x_j+\theta h)\mathrm{d}\theta$$

$$\approx h\int_0^1 [1 + \theta\Delta]f_j\mathrm{d}\theta \qquad = h[(\theta + \tfrac{1}{2}\theta^2\Delta)f_j]_0^1$$

$$= h[1 + \tfrac{1}{2}\Delta]f_j \qquad = h[f_j + \tfrac{1}{2}(f_{j+i} - f_j)]$$

$$= \tfrac{1}{2}h(f_j + f_{j+1}),$$

to obtain the *trapezoidal rule*

$$\int_a^b f(x)\mathrm{d}x \approx \frac{h}{2}(f_0+f_1) + \dots + \frac{h}{2}(f_{N-1}+f_N).$$

$$= \frac{h}{2}(f_0+f_N) + h(f_1+f_2+\dots+f_{N-1}).$$

Integration by the trapezoidal rule therefore involves computing a finite sum of values given by the integrand $f(x)$, and is very quick.

Note that this procedure can be interpreted geometrically (see Figure 15) as the sum of the areas of N trapezoids of width h and average height $\tfrac{1}{2}(f_j+f_{j+1})$.

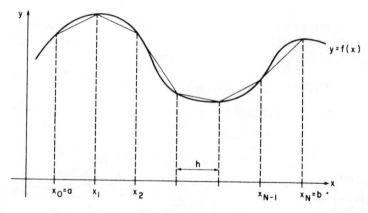

FIGURE 15: The trapezoidal rule.

2   ACCURACY

The trapezoidal rule corresponds to a rather crude polynomial approximation (a straight line) between successive points x_j, $x_{j+1} = x_j + h$, and hence can only be accurate for sufficiently small h. An approximate (upper) bound on the error may be derived as follows. From the Taylor expansion,

$$f_{j+1} = f(x_j + h) = f_j + h f'_j + \frac{1}{2!} h^2 f''_j + \ldots$$

one has the trapezoidal form

$$\int_{x_j}^{x_{j+1}} f(x)dx \approx \tfrac{1}{2} h(f_j + f_{j+1}) \qquad = h(f_j + \frac{h}{2} f'_j + \frac{h^2}{4} f''_j + \ldots),$$

while we may expand $f(x)$ in $x_j \leqslant x \leqslant x_{j+1}$ as

$$f(x) = f_j + (x - x_j)f'_j + \frac{1}{2!}(x - x_j)^2 f''_j + \ldots$$

to get the exact form

$$\int_{x_j}^{x_{j+1}} f(x)dx = h(f_j + \frac{h}{2} f'_j + \frac{h^2}{6} f''_j + \ldots).$$

Comparison of these two forms shows that the correction needed is

$$h(\tfrac{1}{6} - \tfrac{1}{4}) h^2 f''_j + \ldots = -\tfrac{1}{12} h^3 f''_j + \ldots .$$

If we ignore higher order terms, the approximate bound on the error in using the trapezoidal rule (over N sub-intervals) is therefore

$$-\frac{N}{12} h^3 \, \text{Max}\{f''(x)\} = -\frac{(b-a)h^2}{12} \, \text{Max}\{f''(x)\}.$$

Where possible, we choose h small enough to make this error negligible. In the case of desk computation from tables, it may not be possible. On the other hand, in an electronic computer program in which $f(x)$ may be generated anywhere in $a \leqslant x \leqslant b$, the interval may be subdivided smaller and smaller until there is sufficient accuracy. (The integral value for successive sub-divisions can be compared, and the subdivision process terminated when there is adequate agreement between successive values.)

3   EXAMPLE

Estimate $\int_{0\cdot1}^{0\cdot3} e^x \, dx$, using the trapezoidal rule and the data in Step 17 (p. 79).

If we use $T(h)$ to denote the approximation with strip width h we obtain

$$T(0{\cdot}2) = \frac{0{\cdot}2}{2}[1{\cdot}10517 + 1{\cdot}34986] = 0{\cdot}24550$$

$$T(0{\cdot}1) = \frac{0{\cdot}1}{2}[1{\cdot}10517 + 2(1{\cdot}22140) + 1{\cdot}34986]$$
$$= 0{\cdot}24489$$

$$T(0{\cdot}05) = \frac{0{\cdot}05}{2}[1{\cdot}10517 + 2(1{\cdot}16183 + 1{\cdot}22140 + 1{\cdot}28403) + 1{\cdot}34986]$$
$$= 0{\cdot}24474$$

Since $\int_{0{\cdot}1}^{0{\cdot}3} e^x\,dx = e^{0{\cdot}3} - e^{0{\cdot}1} = 0{\cdot}24469$, we may observe that the error sequence $-0{\cdot}00081$, $-0{\cdot}00020$, $-0{\cdot}00005$ corresponds to a decrease in h^2.

Checkpoint

1. Why is quadrature using a polynomial approximation for the integrand likely to be satisfactory, even if the polynomial is of low degree?
2. What is the degree of the approximating polynomial corresponding to the trapezoidal rule?
3. Why is the trapezoidal rule well suited for automatic computation?

EXERCISES

1. Estimate
$$\int_{1{\cdot}00}^{1{\cdot}30} \sqrt{x}\,dx,$$

   using the trapezoidal rule and the data given in the previous Step.
2. Use the trapezoidal rule with $h = 1, 0{\cdot}5$ and $0{\cdot}25$ to estimate
$$\int_{0}^{1} \frac{1}{1+x}\,dx.$$

NUMERICAL INTEGRATION 2
Simpson's rule

If it is undesirable (e.g. in the use of tables) to increasingly sub-divide an interval $a \leqslant x \leqslant b$ in order to get increasingly accurate quadrature, the alternative is to use an approximating polynomial of higher degree. An integration based on quadratic (i.e. parabolic) approximation called *Simpson's rule* is adequate for most quadratures.

1 SIMPSON'S RULE

Simpson's rule corresponds to *quadratic* approximation; thus, for $x_j \leqslant x \leqslant x_j + 2h$,

$$\int_{x_j}^{x_j + 2h} f(x)\mathrm{d}x = h\int_0^2 f(x_j + \theta h)\mathrm{d}\theta$$

$$\approx h\int_0^2 [1 + \theta\Delta + \tfrac{1}{2}\theta(\theta-1)\Delta^2]f_j\mathrm{d}\theta$$

$$= h[(\theta + \tfrac{1}{2}\theta^2\Delta + (\tfrac{1}{6}\theta^3 - \tfrac{1}{4}\theta^2)\Delta^2)f_j]_0^2$$

$$= h[2f_j + 2(f_{j+1}-f_j) + \tfrac{1}{3}(f_{j+2}-2f_{j+1}+f_j)]$$

$$= \tfrac{1}{3}h(f_j + 4f_{j+1} + f_{j+2}).$$

A parabolic arc is fitted to the curve $y = f(x)$ at the *three* tabular points x_j, x_j+h and x_j+2h. Consequently, if $N = (b-a)/h$ is even, one obtains *Simpson's rule*:

$$\int_a^b f(x)\mathrm{d}x = \int_{x_0}^{x_2} f(x)\mathrm{d}x + \int_{x_2}^{x_4} f(x)\mathrm{d}x + \dots + \int_{x_{N-2}}^{x_N} f(x)\mathrm{d}x$$

$$= \tfrac{1}{3}h\,[f_0 + 4f_1 + 2f_2 + 4f_3 + 2f_4 + \dots + 4f_{N-1} + f_N],$$

where

$$f_j = f(x_j) = f(a+jh), \qquad j = 0, 1, 2, ..., N.$$

Integration by Simpson's rule involves computing a finite sum of values given by the integrand $f(x)$, as does the trapezoidal rule. Simpson's rule is also effective for automatic computation, and one direct application in desk calculation usually gives sufficient accuracy.

2   ACCURACY

In automatic computation involving a known integrand $f(x)$, we emphasize that it is quite appropriate to program increased interval subdivision to provide the desired accuracy, but that for desk calculation a (truncation) error bound is again useful.

Suppose that in $x_j - h \leqslant x \leqslant x_j + h$ the function $f(x)$ has the Taylor expansion

$$f(x) = f_j + (x-x_j)f_j' + \frac{1}{2!}(x-x_j)^2 f_j'' + ... \ ;$$

then

$$\int_{x_j-h}^{x_j+h} f(x)dx = 2h\left[f_j + \frac{1}{3}\frac{h^2}{2!}f_j'' + \frac{1}{5}\frac{h^4}{4!}f_j^{(4)} + ...\right].$$

One may re-express the quadrature rule for $x_j - h \leqslant x \leqslant x_j + h$ by $f_{j+1} = f(x_j+h)$ and $f_{j-1} = f(x_j-h)$ as Taylor series; thus

$$\int_{x_j-h}^{x_j+h} f(x)dx \approx \frac{1}{3}h(f_{j-1} + 4f_j + f_{j+1})$$

$$= \frac{1}{3}h\left[(f_j - hf_j' + \frac{1}{2!}h^2f_j'' - ...) + 4f_j\right.$$

$$\left. + (f_j + hf_j' + \frac{1}{2!}h^2f_j'' + ...)\right]$$

$$= 2h(f_j + \frac{1}{3}\frac{h^2}{2!}f_j'' + \frac{1}{3}\frac{h^4}{4!}f_j^{(4)} + ...).$$

Comparison of these two forms shows that the correction needed is

$$2h(\tfrac{1}{5}-\tfrac{1}{3})\frac{h^4}{4!}f_j^{(4)} + ... = -\frac{1}{90}h^5 f_j^{(4)} + ...$$

Ignoring higher order terms, we conclude that the approximate bound on the truncation error in estimating

$$\int_a^b f(x)\mathrm{d}x$$

by Simpson's rule (with $\dfrac{N}{2}$ subintervals of width $2h$) is

$$-\frac{N}{2}\frac{1}{90} h^5 \operatorname{Max}\{f^{(4)}(x)\} = -\frac{(b-a)h^4}{180}.\ \operatorname{Max}\{f^{(4)}(x)\}.$$

It is notable that the error bound is proportional to h^4, compared with h^2 for the cruder trapezoidal rule. In passing, one may note that Simpson's rule is exact for a cubic.

3 . EXAMPLE

Estimate

$$\int_{1\cdot00}^{1\cdot30} \sqrt{x}\ \ \mathrm{d}x$$

using Simpson's rule and the data in Step 26.

There will be an even number of intervals if we choose $h = 0\cdot15$, $0\cdot05$. If we use $S(h)$ to denote the approximation with strip width h, we obtain

$$S(0\cdot15) = \frac{0\cdot15}{3}[1 + 4(1\cdot07238) + 1\cdot14018] = 0\cdot321485,$$

and

$$\begin{aligned} S(0\cdot05) = \frac{0\cdot05}{3}[1 &+ 4(1\cdot02470 + 1\cdot07238 + 1\cdot11803)\\ &+ 2(1\cdot04881 + 1\cdot09545) + 1\cdot14018]\\ = 0\cdot&321486. \end{aligned}$$

Since $f^{(4)}(x) = -\frac{15}{16}x^{-7/2}$, the approximate truncation error bound is

$\dfrac{0\cdot30}{180}\dfrac{15}{16} h^4 = 0\cdot0015625h^4$, whence $0\cdot0000008$ for $h = 0\cdot15$ and $0\cdot00000001$ for $h = 0\cdot05$. Note that the truncation error is negligible; within round-off error, the estimate is $0\cdot32148(6)$.

Checkpoint

---

1. What is the degree of the approximating polynomial corresponding to Simpson's rule?
2. What is the error bound for Simpson's rule?
3. Why is Simpson's rule well suited for automatic computation?

EXERCISE

Estimate

$$\int_0^1 \frac{1}{1+x} \, dx$$

to 4D, using numerical integration.

NUMERICAL INTEGRATION 3
Quadrature from a table of values

In general, formulae for quadrature from a table of finite differences can be obtained by integrating the interpolation formulae. The trapezoidal rule corresponds to linear truncation after integration over $x_j \leqslant x \leqslant x_j + h$, while Simpson's rule corresponds to quadratic truncation after integration over $x_j \leqslant x \leqslant x_j + 2h$. While the trapezoidal rule or Simpson's rule is usually adequate for quadrature, particularly in automatic computation, the integration of a tabular function may require higher order truncation (corresponding to a collocation polynomial of degree higher than 2).

1 QUADRATURE BETWEEN ADJACENT TABULAR POINTS

For integration over an extensive interval, we add formulae for subintervals as before. However, for integration between adjacent tabular points we have to adopt a single interval and retain sufficient terms for accuracy. It is implicit in this procedure that tabular values *outside* the interval of integration are matched by the corresponding higher degree polynomials; such processes are called *partial range integration formulae*. In practice, the point of truncation (polynomial degree) is judged from the size of the higher order difference terms.

i) *Integration of the Newton forward difference formula*

It is sufficient to consider an interval of width h:

$$\int_{x_0}^{x_0+h} f(x)\mathrm{d}x = h\int_0^1 [1 + \theta\Delta + \tfrac{1}{2}\theta(\theta-1)\Delta^2 + \frac{\theta(\theta-1)(\theta-2)}{3!}\Delta^3 + \ldots]f_0 \mathrm{d}\theta$$

$$= h[(\theta + \tfrac{1}{2}\theta^2\Delta + \frac{\tfrac{1}{3}\theta^3 - \tfrac{1}{2}\theta^2}{2!}\Delta^2 + \frac{\tfrac{1}{4}\theta^4 - \theta^3 + \theta^2}{3!}\Delta^3 + \ldots)f_0]_0^1$$

$$= h(f_0 + \tfrac{1}{2}\Delta f_0 - \tfrac{1}{12}\Delta^2 f_0 + \tfrac{1}{24}\Delta^3 f_0 + \ldots).$$

ii) *Integration of the Newton backward difference formula*

$$\int_{x_0}^{x_0+h} f(x)dx = h\int_0^1 \left[1 + \theta\nabla + \tfrac{1}{2}\theta(\theta+1)\nabla^2 + \right.$$
$$\left. + \frac{\theta(\theta+1)(\theta+2)}{3!}\nabla^3 + \ldots\right]f_0 d\theta$$

$$= h(f_0 + \tfrac{1}{2}\nabla f_0 + \tfrac{5}{12}\nabla^2 f_0 + \tfrac{3}{8}\nabla^3 f_0 + \tfrac{251}{720}\nabla^4 f_0 + \ldots).$$

2   EXAMPLE

Evaluate

$$\text{(i)} \int_{1\cdot00}^{1\cdot05} \sqrt{x}\, dx, \qquad \text{and (ii)} \int_{1\cdot25}^{1\cdot30} \sqrt{x}\, dx$$

using the data for $x = 1\cdot00\ (0\cdot05)\ 1\cdot30$ from Step 26.

i) $x_0 = 1\cdot00,\quad x_0+h = 1\cdot05$ (forward differences)

$$\int_{1\cdot00}^{1\cdot05} \sqrt{x}\, dx \approx 0\cdot05\,(1 + 0\cdot01235 + 0\cdot00004(9) + 0\cdot00000(2)\,)$$

$$= 0\cdot05\,(1\cdot01240)$$

$$= 0\cdot05062.$$

ii) $x_0 = 1\cdot25,\quad x_0+h = 1\cdot30$ (backward differences)

$$\int_{1\cdot25}^{1\cdot30} \sqrt{x}\, dx \approx 0\cdot05\,(1\cdot11803 + 0\cdot01129 - 0\cdot00020(4) + 0\cdot00000(4)\,)$$

$$= 0\cdot05\,(1\cdot12912)$$

$$= 0\cdot05646.$$

Checkpoint

1. When may Simpson's rule be inadequate for quadrature?
2. How may the integral of a function be calculated between adjacent tabular points?
3. Which tabular points are used for cubic collocation in quadrature between two adjacent tabular points?

EXERCISES

A function is tabulated as

x	$f(x)$	x	$f(x)$	x	$f(x)$	x	$f(x)$
0·68	0·8087	0·80	1·0296	0·92	1·3133	1·04	1·7036
0·72	0·8771	0·84	1·1156	0·96	1·4284	1·08	1·8712
0·76	0·9505	0·88	1·2097	1·00	1·5574	1·12	2·0660

i) Use forward differences to evaluate

$$\int_{0.88}^{0.92} f(x)\,dx\,;$$

ii) Use backward differences to evaluate

$$\int_{0.88}^{0.92} f(x)\,dx.$$

NUMERICAL INTEGRATION 4
Gauss integration formulae

The numerical integration procedures previously discussed (viz. the trapezoidal rule, Simpson's or higher degree rules from the interpolation formulae) involve equally spaced values of the argument. However, for a fixed number of points the accuracy may be increased if we do not insist that the points are equidistant. This is the background of an alternative integration process due to Gauss, which will now be considered. Briefly, assuming some specified *number* of values of the integrand (of unspecified position) we construct a formula by *deciding* the tabular positions within the range of integration that produce the *most accurate* integration formula.

1 GAUSS TWO-POINT INTEGRATION FORMULA[†]

Consider any two-point formula of form

$$\int_{-1}^{1} f(x)\mathrm{d}x \approx af(\alpha) + bf(\beta),$$

where the *weights* a, b and the *abscissae* α, β are to be determined such that the formula integrates 1, x, x^2, x^3 (and hence all cubic functions) exactly. We have four conditions on the four unknowns, as follows:

i) $f(x) = 1$ integrates exactly if $2 = a + b$;

ii) $f(x) = x$ integrates exactly if $0 = a\alpha + b\beta$;

iii) $f(x) = x^2$ integrates exactly if $\frac{2}{3} = a\alpha^2 + b\beta^2$;

iv) $f(x) = x^3$ integrates exactly if $0 = a\alpha^3 + b\beta^3$.
It is easily verified that

$$a = b = 1, \quad \alpha = -\beta, \quad \alpha^2 = \tfrac{1}{3},$$

and we have the *Gauss two-point integration formula*

---

† This algorithm is suitable for automatic computation. A flow-chart for study and use in programming may be found on page 157.

$$\int_{-1}^{1} f(x)dx \approx f(-\frac{1}{\sqrt{3}}) + f(\frac{1}{\sqrt{3}})$$

$$\approx f(-0{\cdot}57735027) + (f(0{\cdot}57735027).$$

A change of variable also renders this last form applicable to any interval; we make the substitution

$$u = \tfrac{1}{2}[(b-a)x + (b+a)]$$

in the integral we seek to evaluate,

$$\int_{a}^{b} \phi(u)du, \quad \text{say.}$$

If we write

$$\phi(u) = \phi\{\tfrac{1}{2}[(b-a)x + (b+a)]\} \equiv f(x),$$

then

$$\int_{a}^{b} \phi(u)du = \tfrac{1}{2}(b-a) \int_{-1}^{1} f(x)dx,$$

since $du = \tfrac{1}{2}(b-a)dx$, and $u = a$ when $x = -1$, $u = b$ when $x = 1$.

It is important to note that the Gauss two-point formula is exact for cubic polynomials, and hence may be compared in accuracy with Simpson's rule. (In fact, the truncation error for the Gauss formula is about 2/3 that for Simpson's rule.) Since one fewer function value is required for the Gauss formula, it is to be preferred provided the irrationality of the abscissae is unimportant (e.g., in automatic computation).

2 OTHER GAUSS FORMULAE

The Gauss two-point integration formula discussed is but one of a large family of such formulae. Thus, we might derive the *Gauss three-point integration formula*

$$\int_{-1}^{1} f(x)dx \approx \tfrac{1}{9}[5f(-\sqrt{3/5}) + 8f(0) + 5f(\sqrt{3/5})],$$

which is exact for quintics; indeed, the error is less than

$$\frac{1}{15\,750} \text{Max}\{f^{(6)}(x)\}$$

This and the previous two point formula represent the lowest order in a series of formulae commonly referred to as *Gauss-Legendre*, because of their association with the *Legendre polynomials*.

There are yet other formulae associated with other orthogonal polynomials (Laguerre, Hermite, Chebyshev, etc.); the *general form* of Gaussian integration may be represented by the formula

$$\int_a^b w(x)f(x)dx \approx \sum_{i-1}^n A_i f(x_i),$$

where $\{x_1, x_2, ..., x_n\}$ is the set of points in the integration range $a \leqslant x \leqslant b$, the A_i are constants, and $w(x)$ is the so-called *weight function*.

3   APPLICATION OF GAUSS QUADRATURE

In general, the sets $\{x_i\}$, $\{A_i\}$ are tabulated ready for reference, so that application of Gauss quadrature is immediate.

Example

Apply the Gauss (Gauss-Legendre) two-point formula to calculate

$$\int_0^{\pi/2} \sin t \, dt.$$

Recalculate the integral using the Gauss four-point formula.
    The two-point formula $(n = 2)$ is

$$\int_{-1}^1 f(x)dx \approx f(-0.57735027) + f(0.57735027).$$

Change of variable

$$t = \tfrac{1}{2}(\frac{\pi}{2} x + \frac{\pi}{2}) = \pi(x+1)/4,$$

$$\int_0^{\pi/2} \sin t \, dt = \frac{\pi}{4} \int_{-1}^1 \sin \frac{\pi(x+1)}{4} dx$$

$$f(x) = \sin \frac{\pi(x+1)}{4} \Rightarrow f(-0.57735027) = 0.32589$$
$$f(0.57735027) = 0.94541$$

$$\int_0^{\pi/2} \sin t \, dt = \frac{\pi}{4}(0.32589 + 0.94541)$$

$$= 0.99848.$$

The four-point formula ($n = 4$)

$$\int_{-1}^{1} f(x)dx \approx 0.34785485[f(-0.86113631) + f(0.86113631)]$$
$$+ \; 0.65214515[f(-0.33998104) + f(0.33998104)]$$

leads to

$$\int_{0}^{\pi/2} \sin t \, dt = 1.000000,$$

correct to *six* decimal places. This accuracy is impressive enough; the Simpson rule with 64 points produces 0.99999983!

Checkpoint

---

1. What is a disadvantage of integration formulae using constant argument interval?
2. What is the general form of the Gauss integration formula?
3. How accurate are the Gauss-Legendre two-point and three-point formulae?

EXERCISE

Apply the Gauss two-point and four-point formulae to evaluate

$$\int_{0}^{1} \frac{1}{1+u} \, du.$$

NUMERICAL INTEGRATION 5*
Differential equations

In pure mathematics courses a lot of attention is paid to the properties of differential equations and 'analytic' techniques for solving them. Unfortunately, many differential equations (including nearly all the non-linear ones) encountered in the 'real world' are not amenable to analytic solution. Even the apparently simple problem of solving

$$y' = \frac{x+y}{x-y} \qquad \text{with } y = 0 \text{ when } x = 1$$

involves a lot of manipulation before the unwieldy solution

$$\log_e(x^2 + y^2) = 2\tan^{-1}\frac{y}{x}$$

is obtained. Even then a lot more effort is required just to extract the value of y corresponding to *one* value of x. In such situations it is preferable to use a numerical approach from the start.

Partial differential equations are beyond the scope of this text, but we can have a brief look at some methods for solving the single first order *ordinary* differential equation

$$y' = f(x, y)$$

given the *initial value* $y(x_0) = y_0$. Numerical solution involves estimating values of $y(x)$ at (usually equidistant) points $x_1, x_2, ..., x_N$.

1 TAYLOR SERIES

We already have one technique available for this problem; we can estimate $y(x_1)$ by a pth order Taylor series (the particular value of p will depend on the size of $(x_1 - x_0)$ and the accuracy required):

$$y(x_1) \approx y_1 = y(x_0) + (x_1 - x_0)y'(x_0) + \frac{(x_1 - x_0)^2}{2}y''(x_0) + ...$$
$$+ \frac{(x_1 - x_0)^p}{p!}y^{(p)}(x_0).$$

Here, $y(x_0)$ is given, and $y'(x_0)$ can be found by substituting $x = x_0$ and $y = y_0$ in the differential equation, but $y''(x_0)$, ..., $y^{(p)}(x_0)$ require differentiation of $f(x, y)$, which can be messy. Note that $y_1, y_2, ..., y_N$ will be used to denote the _estimates_ of $y(x_1)$, $y(x_2)$, ..., $y(x_N)$.

Once y_1 has been computed, we can estimate $y(x_2)$ by a Taylor Series based either on x_1 (in which case the error in y_1 will be propagated) or on x_0 (in which case p may have to be increased). In the _local_ approach, y_{n+1} is computed from a Taylor Series based on x_n, while in the _global_ approach $y_1, y_2, ..., y_N$ are all computed from Taylor Series based on x_0.

One way of avoiding the differentiation of $f(x, y)$ is to fix $p = 1$ and compute

$$y_{n+1} = y_n + (x_{n+1} - x_n)f(x_n, y_n), \qquad n = 0, 1, ..., N-1.$$

This is known as _Euler's method_. However, unless the distances between the x_i are very small, truncation error will be large and the results inaccurate.

2   RUNGE-KUTTA METHODS[†]

A popular way of avoiding the differentiation of $f(x, y)$ without sacrificing accuracy involves estimating $y(x_{n+1})$ from y_n and a weighted average of values of $f(x, y)$, chosen so that the truncation error is comparable to that of a pth order Taylor Series. The details of the derivation lie beyond the scope of this book but we can quote two of the simpler _Runge-Kutta methods_.

The first has the same order of accuracy as the Taylor series with $p = 2$ and is usually written as three steps:

$$k_1 = h_n f(x_n, y_n);$$
$$k_2 = h_n f(x_n + h_n, y_n + k_1);$$
$$y_{n+1} = y_n + \tfrac{1}{2}(k_1 + k_2).$$

The second is the fourth order method:

$$k_1 = h_n f(x_n, y_n);$$
$$k_2 = h_n f(x_n + \tfrac{1}{2}h_n, y_n + \tfrac{1}{2}k_1);$$
$$k_3 = h_n f(x_n + \tfrac{1}{2}h_n, y_n + \tfrac{1}{2}k_2);$$
$$k_4 = h_n f(x_n + h_n, y_n + k_3);$$
$$y_{n+1} = y_n + \tfrac{1}{6}(k_1 + 2k_2 + 2k_3 + k_4).$$

† This algorithm is suitable for automatic computation. A flow-chart for study and use in programming may be found on page 158.

The *step length* h_n ($=x_{n+1}-x_n$) will be constant if the points are equidistant. Neither method involves evaluating derivatives of $f(x, y)$; instead $f(x, y)$ itself is evaluated several times (twice in the second order method, four times in the fourth).

3 MULTISTEP METHODS

The Taylor series and Runge-Kutta methods are classified as *single step* methods, because the only value of the approximate solution used in constructing y_{n+1} is y_n, the result of the previous step. *Multistep* methods also make use of earlier values like $y_{n-1}, y_{n-2}, ...,$ in order to reduce the number of times $f(x, y)$ or its derivatives have to be evaluated.

Among the multistep methods that can be derived by integrating interpolating polynomials we have (using f_n to denote $f(x_n, y_n)$):

a) the midpoint method (second order):

$$y_{n+1} = y_{n-1} + 2hf_n;$$

b) Milne's method (fourth order):

$$y_{n+1} = y_{n-3} + \tfrac{4}{3} h (2f_n - f_{n-1} + 2f_{n-2});$$

c) the Adams-Bashforth method of order 4:

$$y_{n+1} = y_n + \tfrac{1}{24} h (55f_n - 59f_{n-1} + 37f_{n-2} - 9f_{n-3}).$$

Here the points used in each formula are equally spaced with step-length h. We will not go into the various ways in which multistep methods are used but clearly we will need more than one 'starting value', which is a disadvantage. On the other hand, to construct y_{n+1} we need only evaluate $f(x, y)$ *once*, since $f_{n-1}, f_{n-2}, ...,$ will already have been computed. Which of the three types of methods (Taylor, Runge-Kutta or multistep) is best depends largely on the complexity of $f(x, y)$. Other considerations are discussed in more advanced texts.

4 EXAMPLE

It is instructive to compare some of the methods given above on a very simple problem: estimate $y(0.5)$, given that

$$y' = x + y \quad \text{with } y(0) = 1, \quad \text{i.e. } x_0 = 0, y_0 = 1.$$

The exact solution is

$$y(x) = 2e^x - x - 1$$

so

$$y(0.5) = 1.79744.$$

We shall use a fixed step length $h = 0.1$ and work to 5D.

a)   Euler's method (first order):

$$y_{n+1} = y_n + 0.1(x_n + y_n) = 0.1x_n + 1.1y_n$$

so

$$
\begin{aligned}
y_1 &= 0.1(0) + 1.1(1) = 1.1 \\
y_2 &= 0.1(.1) + 1.1(1.1) = 1.22 \\
y_3 &= 0.1(.2) + 1.1(1.22) = 1.362 \\
y_4 &= 0.1(.3) + 1.1(1.362) = 1.5282
\end{aligned}
$$

and $y_5 = 0.1(.4) + 1.1(1.5282) = 1.72102$

which has an error of approximately 0.08.

b)   Taylor Series (fourth order):

Since

$$y' = x+y, \quad y'' = 1+y', \quad y''' = y'' \quad \text{and} \quad y^{(4)} = y''',$$

we have

$$y_{n+1} = y_n + 0.1(x_n + y_n) + \frac{(0.1)^2}{2}(1 + x_n + y_n) + \frac{(0.1)^3}{6}(1 + x_n + y_n)$$
$$+ \frac{(0.1)^4}{24}(1 + x_n + y_n)$$

$$\approx 0.00517 + 0.10517x_n + 1.10517y_n.$$

Thus

$$
\begin{aligned}
y_1 &= 0.00517 + 0.10517(0) + 1.10517(1) = 1.11034 \\
y_2 &= 0.00517 + 0.10517(0.1) + 1.10517(1.11034) = 1.24280 \\
y_3 &= 0.00517 + 0.10517(0.2) + 1.10517(1.24280) = 1.39971 \\
y_4 &= 0.00517 + 0.10517(0.3) + 1.10517(1.39971) = 1.58364
\end{aligned}
$$

and $y_5 = 0.00517 + 0.10517(0.4) + 1.10517(1.58364) = 1.79743$

which has an error of 0.00001.

c)   Runge-Kutta (second order):

$k_1 = 0.1(x_n + y_n)$, $k_2 = 0.1(x_n + 0.1 + y_n + k_1)$, $y_{n+1} = y_n + \frac{1}{2}(k_1 + k_2)$

$n = 0$: $k_1 = 0.1(0+1) = 0.1$, $k_2 = 0.1(0.1 + 1.1) = 0.12$, $y_1 = 1 + \frac{1}{2}(0.1 + 0.12)$
$$= 1.11$$

$n = 1$: $k_1 = 0.1(0.1 + 1.11)$, $k_2 = 0.1(0.2 + 1.11 + 0.121)$,
$$y_2 = 1.11 + \frac{1}{2}(0.121 + 0.1431) = 1.24205$$

$n = 2$: $k_1 = 0.1(0.2 + 1.24205)$, $k_2 = 0.1(0.3 + 1.24205 + 0.14421)$
$$y_3 = 1.24205 + \frac{1}{2}(0.14421 + 0.16863) = 1.39847$$

$n = 3$: $k_1 = 0.1(0.3 + 1.39847)$, $k_2 = 0.1(0.4 + 1.39847 + 0.16985)$
$$y_4 = 1.39847 + \frac{1}{2}(0.16985 + 0.19683) = 1.58181$$

$n = 4 : k_1 = 0\cdot1(0\cdot4 + 1\cdot58181), \; k_2 = 0\cdot1(0\cdot5 + 1\cdot58181 + 0\cdot19818),$
$$y_5 = 1\cdot58181 + \tfrac{1}{2}(0\cdot19818 + 0\cdot22800) = 1\cdot79490$$

y_5 is in error by approximately $0\cdot003$.

As we might expect, the fourth order method is clearly superior, the first order method is clearly inferior, and the second order method falls in between.

Checkpoint

---

1. For each of the three types of method outlined in this Step, what is the main disadvantage?
2. Why might we expect higher order methods to be more accurate?

EXERCISES

1. Use the midpoint method with step length $h = 0\cdot1$ to estimate $y(0\cdot5)$ given that $y' = x + y$ with $y(0) = 1$.
   Take $y_1 = 1\cdot11$ (from (c) above) as the second starting value.
2. Use Euler's method with step length $h = 0\cdot2$ to estimate $y(1)$ given that $y' = -xy^2$ with $y(0) = 2$.

APPENDIX
Flow-charts

Basic flow-charts are given for some of the algorithms introduced in the book. Students will gain much if they study the flow-chart of a method at the same time as they learn it in a Step. If they are familiar with a programming language they should attempt to convert at least some of the flow-charts into computer programs, and apply them to the set exercises.

CONTENTS

1 THE BISECTION METHOD (STEP 7)

Assume the equation is $f(x) = 0$.

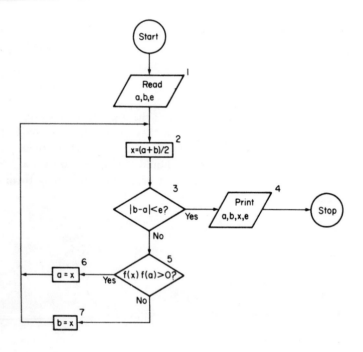

Points for study

a) What are the input values used for?
b) Explain the purpose of blocks 5, 6 and 7.
c) Amend the flow-chart so that the process will always stop after exactly *M* steps.
d) Amend the flow-chart so that the process will stop as soon as $|f(x)| < e$.
e) Write a computer program for the algorithm.
f) Use the computer program to solve the equation $x + \cos x = 0$.

2   THE METHOD OF FALSE POSITION (STEP 8)

Assume the equation is $f(x) = 0$.

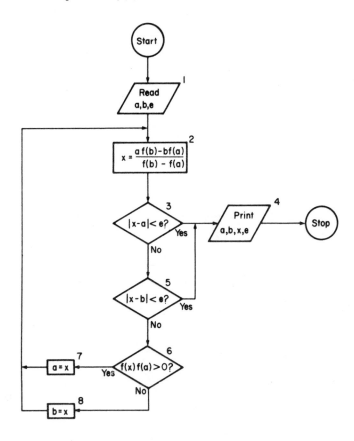

Points for study

a)   What are the input values used for?
b)   Explain the purpose of blocks 3 and 5.
c)   Under what circumstances may the process stop with a large error in x?
d)   Amend the flow-chart so that the process will stop as soon as $|f(x)| < e$.
e)   Write a computer program for the algorithm.
f)   Use the computer program to solve the equation $x + \cos x = 0$.

3   THE NEWTON-RAPHSON METHOD (STEP 10)

Assume the equation is $f(x) = 0$.

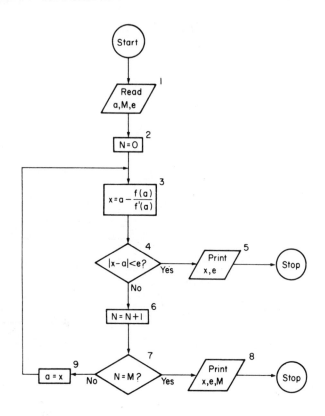

Points for study

a)   What are the input values used for?
b)   Why is M included in the output of block 8?
c)   What happens if $f'(a)$ is very small?
d)   Amend the flow-chart to take suitable action if $f(a)$ is very small.
e)   Write a computer program for the algorithm.
f)   Use the computer program to solve the equation $x + \cos x = 0$.

4   GAUSS-SEIDEL ITERATION (STEP 13)

Assume the system is:

$$a_{11}x + a_{12}y + a_{13}z = b_1,$$
$$a_{21}x + a_{22}y + a_{23}z = b_2,$$
$$a_{31}x + a_{32}y + a_{33}z = b_3.$$

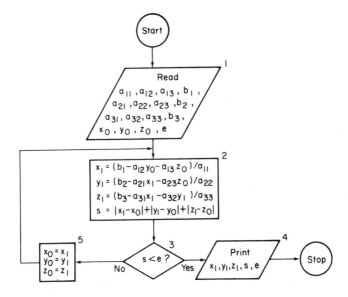

Points for study

a) What are the input values used for?
b) What is the purpose of the number s?
c) Amend the flow-chart so that there is a safeguard against divergence or slow convergence (c.f. flow-chart 3).
d) Amend the flow-chart to solve a general $n \times n$ system $Ax = b$.
e) Write a computer program for the algorithm.
f) Use the computer program to solve the system:

$$8x + y - 2z = 5,$$
$$x - 7y + z = 9,$$
$$2x + \quad 9z = 11.$$

5 LAGRANGE INTERPOLATION (STEP 22)

We require $y(x)$, given that $y_1 = y(x_1)$, $y_2 = y(x_2)$, ..., $y_n = y(x_n)$.

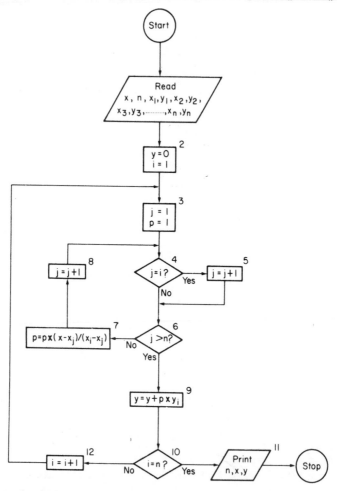

Points for study

a) What is the purpose of blocks 4 and 5?
b) What value (in algebraic terms) does p have in block 9?
c) What value (in algebraic terms) does y have after block 9?
d) Follow the flow-chart through with the data $x = 1\cdot5$, $n = 3$, $x_1 = 0$, $y_1 = 2\cdot5$, $x_2 = 1$, $y_2 = 4\cdot7$, $x_3 = 3$, $y_3 = 3\cdot1$.
e) Write a computer program for the algorithm.
f) Use the computer program on the data of (d) above.

6   THE TRAPEZOIDAL RULE (STEP 27)

Assume the integral is $\int_a^b f(x)\mathrm{d}x$.

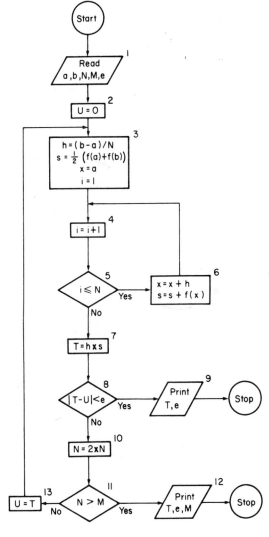

Points for study

a)   What are the input values used for?
b)   What value (in algebraic terms) does T have after block 7?

c) What is the purpose of blocks 8, 10 and 11?
d) Why is M included in the output of block 12?
e) Write a computer program for the algorithm.
f) Use the computer program to evaluate the integrals

$$\int_0^2 \frac{dx}{1 + x^3} \text{ and } \int_0^1 e^{-x^2} \, dx.$$

7   GAUSS INTEGRATION FORMULAE (STEP 30)

Assume the integral is $\int_a^b \phi(u)du$. Use the two-point Gauss formula.

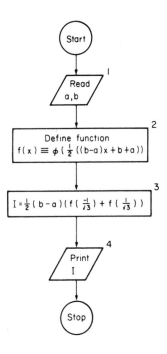

Points for study

a)   What is the purpose of block 2?

b)   What changes are required to produce an algorithm based on the *three* point Gauss formula?

c)   Amend the flow-chart to use successive subdivision of the interval (c.f. flow-chart 6) to evaluate the integral to within a tolerance of *e*.

d)   Write a computer program for the algorithm (as amended in part (c)).

e)   Use the computer program to evaluate the integrals.

$$\int_0^2 \frac{dx}{1+x^3} \qquad \text{and} \int_0^1 e^{-x^2}dx.$$

8   RUNGE-KUTTA METHOD (STEP 31*)

Assume the equation is $\dfrac{dy}{dx} = f(x, y)$.   Use the usual fourth order method.

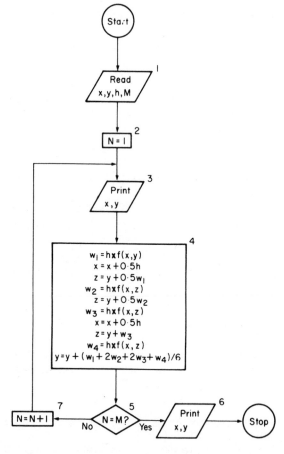

Points for study

a)   What are the input values used for?

b)   How many times is the function f evaluated between block 2 and block 6?

c)   Amend the flow-chart to use the second order Runge-Kutta method.

d)   Write a computer program for the algorithm.

e)   Use the computer program to solve the equation $y' = x - y$, starting with $x = 0$, $y = 1$ and taking 10 steps of length 0·1.

BIBLIOGRAPHY

The following is a short list of books which may be referred to for complementary reading, proofs omitted in this text, or further study in Numerical Analysis.

Calculus
Thomas, *Calculus and Analytic Geometry* (Addison-Wesley, 1967).
Computer Science
Davis, *Introduction to Electronic Computers* (McGraw-Hill, 1971).
Forsythe, Keenan, Organick and Stenberg, *Computer Science* (Wiley, 1969).
Linear Algebra
Murdoch, *Linear Algebra for Undergraduates* (Wiley, 1962).
Numerical Analysis
Acton, *Numerical Methods that work* (Harper and Row, 1970).
Conte and de Boor, *Elementary Numerical Analysis* (McGraw-Hill, 1972).
Faddeeva, *Computational Methods of Linear Algebra* (Dover, 1959).
Fox and Mayers, *Computing Methods for Scientists and Engineers* (Oxford University Press, 1968).

Tables
Abramowitz and Stegun, *Handbook of Mathematical Functions* (Dover, 1965).

Answers to the Exercises

STEP 1

Error points are indicated by asterisks. Note that errors of types (a) and (b) are involved in the use of both formulae.

1. $T \approx 2 \times 3 \cdot 14 \times \sqrt{75/981} \approx 6 \cdot 28 \times \sqrt{0 \cdot 0765} \approx 6 \cdot 28 \times 0 \cdot 277$
    $ * \quad * \quad * * *$

    $\approx 1 \cdot 74$ sec
    $ *$

2. $R \approx 0 \cdot 028 \times 3 \cdot 14 \times 56 \cdot 25 \times \sqrt{2 \times 981 \times 650} \approx 4 \cdot 946 \times 1129$
    $ * * * * * * *$

    $\approx 5 \cdot 58 \times 10^3$ cm^3/sec.
    $ *$

STEP 2

1. $1 \cdot 2345 \times 10^1$, $8 \cdot 0059 \times 10^{-1}$, $2 \cdot 96844 \times 10^2$, $5 \cdot 19 \times 10^{-3}$.

2. a)   34·7,   3·47,   0·347,   0·0347.
    b)   34·782,   3·478,   0·347,   0·034.
    c)   34·8,   3·48,   0·348,   0·0348.
    d)   34·782,   3·478,   0·348,   0·035.

STEP 3

a) The result 13·57,   Max$|e_{abs}| = 0 \cdot 005 + 0 \cdot 005 = 0 \cdot 01$,
    so the answer is $13 \cdot 57 \pm 0 \cdot 01$ or 13·6 correct to 3S.

b) The result 0·01,   Max$|e_{abs}| = 0 \cdot 01$, so that although operands are
    correct to 5S, the answer may not even be correct to 1S! This
    phenomenon is known as *loss of significance* or *cancellation*.

c) The result 13·3651,   Max $|e_{abs}| \approx (4 \cdot 27 + 3 \cdot 13) \times 0 \cdot 005 = 0 \cdot 037$, so
    the answer is $13 \cdot 3651 \pm 0 \cdot 037$ or 13 correct to 2S.

d) The result $- 1.85676$, Max $|e_{abs}| \approx 0.513 \times 0.005 + 9.48 \times 0.0005$ $+ 0.005 \approx 0.012$, so the answer is $- 1.85676 \pm 0.012$ or -2 correct to 1S.

e) The result $1.109...$, $\text{Max}|e_{rel}| \approx \dfrac{\text{Max}|e_{abs}|}{1.109} \approx \dfrac{0.005}{0.25} + \dfrac{0.005}{2.84} + \dfrac{0.005}{0.64}$ ≈ 0.030, so the answer is 1.109 ± 0.033 or 1.1 correct to 2S.

f) The result 0.47, $\text{Max}|e_{abs}| = 7 \times 0.005 = 0.035$, so the answer is 0.47 ± 0.035 and we cannot even guarantee 1S.

STEP 4

1. a) $12.01 \times 10^2 \rightarrow 1.20 \times 10^3$.

   b) $6.19 \times 10^2 + 0.361 \times 10^2 = 6.551 \times 10^2 \rightarrow 6.55 \times 10^2$.

   c) $0.37 \times 10^2 \rightarrow 3.70 \times 10^1$.

   d) $6.19 \times 10^2 - 0.361 \times 10^2 = 5.829 \times 10^2 \rightarrow 5.83 \times 10^2$.

   e) $3.63600 \times 10^2 \rightarrow 3.64 \times 10^2$.

   f) $33.3000 \times 10^0 \rightarrow 3.33 \times 10^1$.

   g) $1.25000 \times 10^3 \rightarrow 1.25 \times 10^3$.

   h) $- 0.869300..., \times 10^{-5} \rightarrow - 8.69 \times 10^{-6}$.

2. Since the initial errors are of unknown sign and size, we estimate E, the maximum *magnitude* of the accumulated error, from the results of Step 3, assuming the worst about the initial errors. To estimate the propagated error we use $\text{Max}|e_{abs}| = |e_1| + |e_2|$ for addition and subtraction, and $\text{Max}|e_{rel}| \approx \left|\dfrac{e_1}{x^*}\right| + \left|\dfrac{e_2}{y^*}\right|$ for multiplication and division. The magnitude of the generated error is denoted by ε.

   a) $\varepsilon = 0.001 \times 10^3$, $\text{Max}|e_{abs}| = 0.005 \times 10^2 + 0.005 \times 10^2$ $= 0.01 \times 10^2$, $E = 0.002 \times 10^3$.

   b) $\varepsilon = 0.001 \times 10^2$, $\text{Max}|e_{abs}| = 0.005 \times 10^2 + 0.005 \times 10^1$ $= 0.0055 \times 10^2$, $E = 0.0065 \times 10^2$.

   c) $\varepsilon = 0$, $\text{Max}|e_{abs}| = 0.005 \times 10^2 + 0.005 \times 10^2 = 0.01 \times 10^2$, $E = 0.1 \times 10^1$ (relatively large).

   d) as for (b)

e) $\varepsilon = 0.004 \times 10^2$, $\text{Max}|e_{rel}| \approx \dfrac{0.005}{3.60.} + \dfrac{0.005}{1.01}$,

$\text{Max}|e_{abs}| \approx 0.005 \times (1.01 + 3.60) \times 10^2 \approx 0.023 \times 10^2$,
$E \approx 0.027 \times 10^2$.

f) $\varepsilon = 0$, $\text{Max}|e_{abs}| \approx 0.005 \times (7.50 + 4.44) \times 10^0 \approx 0.06 \times 10^0$
$E \approx 0.006 \times 10^1$.

g) $\varepsilon = 0$, $\text{Max}|e_{rel}| \approx \dfrac{0.005}{6.45} + \dfrac{0.005}{5.16}$,

$\text{Max}|e_{abs}| \approx \dfrac{0.005 \times 11.61}{6.45 \times 5.16} \times 1.25 \times 10^3 \approx 0.002 \times 10^3$,

$E \approx 0.002 \times 10^3$.

h) $\varepsilon \approx 0.003 \times 10^{-6}$, $\text{Max}|e_{rel}| \approx \dfrac{0.005}{2.86} + \dfrac{0.005}{3.29}$,

$\text{Max}|e_{abs}| \approx \dfrac{0.005 \times 6.15}{2.86 \times 3.29} \times 8.69 \times 10^{-6} \approx 0.028 \times 10^{-6}$,

$E \approx 0.031 \times 10^{-6}$.

3. a) $b - c = 5.685 \times 10^1 - 5.641 \times 10^1 = 0.044 \times 10^1 \rightarrow 4.400 \times 10^{-1}$.
$a(b-c) = 6.842 \times 10^{-1} \times 4.400 \times 10^{-1} = 30.1048 \times 10^{-2}$
$\qquad\qquad\qquad\qquad\qquad\qquad\qquad\qquad \rightarrow 3.010 \times 10^{-1}$.
$ab = 6.842 \times 10^{-1} \times 5.685 \times 10^1 = 38.896770 \times 10^0$
$\qquad\qquad\qquad\qquad\qquad\qquad\qquad \rightarrow 3.890 \times 10^1$.
$ac = 6.842 \times 10^{-1} \times 5.641 \times 10^1 = 38.595722 \times 10^0$
$\qquad\qquad\qquad\qquad\qquad\qquad\qquad \rightarrow 3.860 \times 10^1$.
$ab - ac = 3.890 \times 10^1 - 3.860 \times 10^1 = 0.030 \times 10^1$
$\qquad\qquad\qquad\qquad\qquad\qquad\qquad \rightarrow 3.000 \times 10^{-1}$.

The answer obtained (working to 6S) is 3.01048×10^{-1}, with propagated error at most 0.069×10^{-1} (so we can only rely on the first figure!).

b) $a + b = 9.812 \times 10^1 + 0.04631 \times 10^1 = 9.85831 \times 10^1$
$\qquad\qquad\qquad\qquad\qquad\qquad\qquad\qquad \rightarrow 9.858 \times 10^1$.
$(a + b) + c = 9.858 \times 10^1 + 0.08340 \times 10^1 = 9.94140 \times 10^1$
$\qquad\qquad\qquad\qquad\qquad\qquad\qquad\qquad \rightarrow 9.941 \times 10^1$.
$b + c = 4.631 \times 10^{-1} + 8.340 \times 10^{-1} = 12.971 \times 10^{-1}$
$\qquad\qquad\qquad\qquad\qquad\qquad\qquad\qquad \rightarrow 1.297 \times 10^0$.
$a + (b + c) = 9.812 \times 10^1 + 0.1297 \times 10^1 = 9.9417 \times 10^1$
$\qquad\qquad\qquad\qquad\qquad\qquad\qquad\qquad \rightarrow 9.942 \times 10^1$.

The answer obtained (working to 6S) is 9.94171×10^1, with propagated error at most 0.00051×10^1.

STEP 5

1. $f(x) = \cos x, \ f'(x) = -\sin x, \ f''(x) = -\cos x, \ldots$

$$\cos x = \cos 0 - x \sin 0 - \frac{x^2}{2!} \cos 0 + \ldots$$

$$= 1 - \frac{x^2}{2!} + \frac{x^4}{4!} - \ldots .$$

2. i)   linear: $1 + x$ over the range $-0.1 < x < 0.1$.

   ii)   quadratic: $1 + x + \frac{1}{2} x^2$ over the range $-0.3 < x < 0.3$.

   iii) cubic: $1 + x + \frac{1}{2} x^2 + \frac{1}{6} x^3$ over the range $-0.5 < x < 0.5$.

3. We have

$$e^x \approx 1 + x + \frac{x^2}{2!} + \ldots + \frac{x^n}{n!} + R_n,$$

where

$$R_n = \frac{x^{n+1}}{(n+1)!} e^{\xi} < \frac{1^{n+1} e^1}{(n+1)!} \text{ for all } x \text{ between 0 and 1.}$$

Thus $R_n < \frac{1}{2} \times 10^{-5}$ if $(n+1)! \geqslant 2e \times 10^5 \approx 543656$; i.e. we

require $n = 9$ (thus, 10 terms), since $9! = 362880$.

4. $p_0 = 1$, $q_0 = 0$
    $p_1 = p_0(3.1) + (-2) = 1.1$, $q_1 = q_0(3.1) + (1) = 1$
    $p_2 = p_1(3.1) + (2) = 5.41$, $q_2 = q_1(3.1) + (1.1) = 4.2$
    $p_3 = p_2(3.1) + (3) = 19.771$, $q_3 = q_2(3.1) + (5.41) = 18.43$
    Only 3 multiplications and 3 additions are required to evaluate
    $P(3.1)$ whereas $3.1 \times 3.1 \times 3.1 - 2 \times 3.1 \times 3.1 + 2 \times 3.1 + 3$
    requires 5 multiplications and 3 additions.

STEP 6

The curves are sketched in Fig. 16 (on p. 164) in a manner similar to those in Section 2(a) of Step 6, and enable us to deduce that there is one real root near $x = -0.7$. Tabulating confirms this:

x	-0.7	-0.8	-0.75
$\cos x$	0.7648	0.6967	0.7317
$x + \cos x$	0.0648	-0.1033	-0.0183

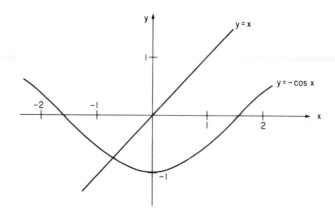

FIGURE 16. Graphs of $y = x$ and $y = -\cos x$

STEP 7

1. In Step 6 we saw that the root lies in the interval $(-0.75, -0.7)$. Successive bisections produce the following sequence of intervals containing the root: $(-0.75, -0.725)$, $(-0.75, -0.7375)$, $(-0.74375, -0.7375)$.

   Thus the root is -0.74 to 2D.

2. Root is 0·615 to 3D.

STEP 8

1. Tabulate $f(x)$:

x	$f(x)$
0	-2
0·2	$-1·4$
0·6	$-0·27$
0·8	$+0·23$

There is a root in the interval $0·6 < x < 0·8$.
We have

$$x = \frac{1}{0·23 + 0·27} \begin{vmatrix} 0·6 & -0·27 \\ 0·8 & 0·23 \end{vmatrix}$$

$$= \frac{0·138 + 0·216}{0·50} = 0·708 \,,$$

$$f(x) = f(0.708)$$
$$= 2\sin(0.708) + 0.708 - 2$$
$$= 1.3006 + 0.708 - 2$$
$$= 0.0086.$$

Since $f(0.6)$ and $f(0.708)$ have opposite signs, the root is in the interval $0.6 < x < 0.78$. Repeating the process,

$$\bar{x} = \frac{1}{0.0086 + 0.27} \begin{vmatrix} 0.6 & -0.27 \\ 0.708 & 0.0086 \end{vmatrix}$$

$$= \frac{0.00516 + 0.19116}{0.2786} = 0.7047 .$$

Since $f(\bar{x}) = f(0.7047) = 0.0003$, we know the root lies between 0.6 and 0.7047, so we compute

$$\bar{x} = \frac{1}{0.0003 + 0.27} \begin{vmatrix} 0.6 & -0.27 \\ 0.7047 & 0.0003 \end{vmatrix} = 0.7046.$$

Repeating the process once more gives 0.7046 (to 4D) again, so we can certainly take 0.705 as our answer. Note that all $\bar{x}$ computed have $f(\bar{x})$ positive.

2. Let us take $f(x) = 3\sin x - x - \frac{1}{x}$. We note that $f(0.7) = -0.1959$ and $f(0.9) = 0.3389$, i.e. the root is enclosed. Working to 4 decimal places, the following results are obtained.

   a) Bisection gives the sequence of intervals
      $(0.7, 0.9)$, $(0.7, 0.8)$, $(0.75, 0.8)$, $(0.75, 0.775)$,
      $(0.7625, 0.775)$, $(0.7625, 0.7688)$, $(0.7625, 0.7657)$,
      $(0.7625, 0.7641)$, $(0.7625, 0.7633)$, $(0.7629, 0.7633)$.
      Since $f(0.7631) = 0.0000$ the process is then terminated.

   b) False position with $a = 0.7$ and $b = 0.9$ gives $\bar{x} = 0.7733$. Since $f(0.7733) = +0.0290$, the process is repeated with $a = 0.7$, $b = 0.7733$ to give $x = 0.7638$. Taking $a = 0.7$, $b = 0.7638$ (since $f(0.7638) = +0.0020$), gives $\bar{x} = 0.7632$ and $f(\bar{x}) = +0.0002$. Finally $a = 0.7$, $b = 0.7632$ gives $\bar{x} = 0.7631$. Note that all $f(\bar{x})$ are positive.

   c) The secant method with $x_0 = 0.7$, $x_1 = 0.9$ gives $x_2 = 0.7733$, $x_3 = 0.7614$ and $x_4 = 0.7631$. This is clearly the most efficient method of the three for this example.

3. Let us take $f(x) = x + \cos x$. In Step 7 we found that the root lies
    in the interval $(-0.74375, -0.7375)$. False position with $a = -0.75$,
    $b = -0.73$ (using $f(a) = -0.01831$, $f(b) = 0.01517$) gives $\bar{x} =$
    -0.73906. Since $f(-0.73906) = 0.00004$, the process is repeated
    with $a = -0.75$, $b = -0.73906$ to give $\bar{x} = -0.73908$. Since
    $f(-0.73908) = 0.00001$, we can give the root as -0.7391 (to 4D).

STEP 9

1. Using the iteration formula

    $$x_{n+1} = 0.5 + \sin x_n,$$

    only six iterations are required:
    $$x_1 = 1.34147,$$
    $$x_2 = 1.47382,$$
    $$x_3 = 1.49530,$$
    $$x_4 = 1.49715$$
    $$x_5 = 1.49729,$$
    $$x_6 = 1.49730.$$

Note that $\phi'(x) = \cos x \approx 0.07$ near the root, so convergence is fast
(and 'one-sided').

2. In Step 7 we found that the root is -0.74 to 2D. Using the iteration
    formula
    $$x_{n+1} = -\cos x_n$$
    with $x_0 = -0.74$, we obtain
    $$x_1 = -0.73847,$$
    $$x_2 = -0.73950,$$
    $$x_3 = -0.73881,$$
    $$x_4 = -0.73927,$$
    $$x_5 = -0.73896,$$
    $$x_6 = -0.73917,$$
    $$x_7 = -0.73903,$$
    $$x_8 = -0.73912,$$
    $$x_9 = -0.73906,$$
    $$x_{10} = -0.73910,$$

Since x_9 and x_{10} agree to 4D we can give the root as -0.7391.
Note that $\phi'(x) = \sin x \approx -0.67$ near the root, so convergence is slow
(and 'oscillatory').

STEP 10

1. Since $x > 0$, the root of $f(x) \equiv \log_e 3x + x = 0$ must lie in the interval $0 < x < \frac{1}{3}$, where $\log_e 3x < 0$. If $x_0 = 0.25$ is the initial guess,

$$f(0.25) = \log_e (0.75) + 0.25$$
$$= -0.2877 + 0.25$$
$$= -0.0377.$$

Since

$$f'(x) = \frac{1}{x} + 1,$$

$$f'(0.25) = 5$$

and

$$x_1 = 0.25 + \frac{0.0377}{5}.$$
$$= 0.25 + 0.0075$$
$$= 0.2575.$$

Then

$$f(0.2575) = \log_e (0.7725) + 0.2575$$
$$= -0.2581 + 0.2575$$
$$= -0.0006$$
$$f'(0.2575) = 3.883 + 1 = 4.883,$$

and

$$x_2 = 0.2575 + \frac{0.0006}{4.883}$$
$$= 0.2575 + 0.0001$$
$$= 0.2576.$$

Since $f(0.2576) = -0.0001$, we conclude that the root is 0.2576 to 45. Note that only 2 or 3 steps are required for the Newton-Raphson process, whereas 8 steps were needed for the iteration method based on

$$x_{n+1} = \frac{1}{3} e^{-x_n}$$

2.
$$x^k = a$$
$$f(x) = x^k - a = 0$$

$$f'(x) = kx^{k-1}$$

$$x_{n+1} = x_n - \frac{x_n^k - a}{kx_n^{k-1}}.$$

With $k = -1$ we have an iterative formula for computing inverses without division: $x_{n+1} = x_n(2 - ax_n)$.

3. $x_0 = 1, a = 10.$

$$x_1 = \frac{1}{2}(1 + \frac{10}{1}) = 5.5,$$

$$x_2 = \frac{1}{2}(5.5 + \frac{10}{5.5}) = 3.66,$$

$$x_3 = \frac{1}{2}(3.66 + \frac{10}{3.66}) = 3.196,$$

$$x_4 = \frac{1}{2}(3.196 + \frac{10}{3.196}) = 3.1625,$$

$$x_5 = \frac{1}{2}(3.1625 + \frac{10}{3.1625}) = 3.16228,$$

$$x_6 = \frac{1}{2}(3.16228 + \frac{10}{3.16228}) = 3.16228.$$

Thus $\sqrt{10}$ is 3.1623 to $4D$.

4. In Step 7 we found that the root is -0.74 to $2D$. Taking $x_0 = -0.74$ and $f(x) = x + \cos x$ so that $f'(x) = 1 - \sin x$,

we obtain $x_1 = -0.74 - \dfrac{(-0.00153)}{1.67429} = -0.73909$

and $x_2 = -0.73909 - \dfrac{(-0.00001)}{1.67361} = -0.73908.$

Since x_1 and x_2 agree to $4D$ we can give the root as -0.7391.

STEP 11

Full answers are given for questions 1 and 2 only.

1.

m	Augmented Matrix				Check
	1	1	-1	0	1
	2	-1	1	6	8
	3	2	-4	-4	-3
	1	1	-1	0	1
-2		-3	3	6	6
-3		-1	-1	-4	-6
	1	1	-1	0	1
		-3	3	6	6
$-1/3$			-2	-6	-8

Solution by back-substitution   *Residuals*

$$\left. \begin{aligned} -2x_3 &= -6 \rightarrow x_3 = 3 \\ -3x_2 + 9 &= 6 \rightarrow x_2 = 1 \\ x_1 + 1 - 3 &= 0 \rightarrow x_1 = 2 \end{aligned} \right\} \quad \begin{aligned} 0 - (2 + 1 - 3) &= 0 \\ 6 - (4 - 1 + 3) &= 0 \\ -4 - (6 + 2 - 12) &= 0 \end{aligned}$$

2.

m	Augmented Matrix			Check	
	5·6	3·8	1·2	1·4	12·0
	3·1	7·1	$-4·7$	5·1	10·6
	1·4	$-3·4$	8·3	2·4	8·7
	5·6	3·8	1·2	1·4	12·0	
$-0·554$		4·99	$-5·36$	4·32	3·95	Working
$-0·250$		$-4·35$	8·00	2·05	5·70	
	5·6	3·8	1·2	1·4	12·0	to 2D
		4·99	$-5·36$	4·32	3·95	
$+0·872$			3·33	5·82	9·14 (9·15)	(rounded)

Solution by back-substitution

$$\left. \begin{aligned} 3·33z &= 5·83 \rightarrow z = 1·75 \\ 4·99y - 5·36 \times 1·75 &= 4·32 \rightarrow y = 2·75 \\ 5·6x + 3·8 \times 2·75 + 1·2 \times 1·75 &= 1·4 \rightarrow x = -1·99 \end{aligned} \right\}$$

Residuals

$$1·4 - (-11·14 + 10·45 + 2·10) = -0·01$$
$$5·1 - (-6·17 + 19·53 - 8·23) = -0·03$$
$$2·4 - (-2·79 - 9·35 + 14·53) = 0·01$$

3. i) $x = 25·5$, $y = -9$, $z = 2$;   *residuals* 0, 0, 0.

ii) $x = -4·30$, $y = -2·42$, $z = 5·07$;   *residuals* 0·00, 0·06, $-0·01$, (2D working),

STEP 12

1. If there were no uncertainties in the constants, the exact solution
    would be $x = 2.6$, $y = 1.2$. With uncertainties the range of solu-
    tions is $2.57 \leqslant x \leqslant 2.63$, $1.17 \leqslant y \leqslant 1.23$.

2. i) $x = 1.2$, $y = 2.3$;
    ii) $x = 1$, $y = 1$, $z = 2$;
    iii) $x_1 = 1.2$, $x_2 = 3.5$, $x_3 = 6.4$.

3. Without pivotal condensation, $x = 1.004$; $y = 0.4998$. With pivotal
    condensation, $x = 1.000$; $y = 0.5000$.

STEP 13

1. $S_3 = |x_1^{(4)} - x_1^{(3)}| + |x_2^{(4)} - x_2^{(3)}| + |x_3^{(4)} - x_3^{(3)}|$.

The values for the fourth iteration $x^{(4)}$ are obtained by continuing the
table of Section 3 for one more row, thus:

4	0·999917	0·999993	1·000017

Then S_3 is found to be:

$$S_3 = 0.001226 + 0.000304 + 0.000215 = 0.001745$$

2. i)   *Rearranged equations:* $20x + 3y - 2z = 51$
         (to place largest coefficients $2x + 8y + 4z = 25$
         on the leading diagonal) $x - y + 10z = -7$

         Recurrence relations: $x^{(k+1)} = 2.55 - 0.15y^{(k)} + 0.1z^{(k)}$

         $y^{(k+1)} = 3.125 - 0.25x^{(k+1)} - 0.5z^{(k)}$

         $z^{(k+1)} = -0.7 - 0.1x^{(k+1)} + 0.1y^{(k+1)}$

Taking as stopping criterion $S_k < 0.00005$, we obtain the following
results.

Iteration k	$x^{(k)}$	$y^{(k)}$	$z^{(k)}$	S_k(to 5D)
0	0	0	0	5·74400
1	2·550	2·488	− 0·706	0·99900
2	2·106	2·952	− 0·615	0·09370
3	2·0457	2·9211	− 0·6125	0·00809
4	2·05059	2·91860	− 0·61320	0·00058
5	2·050890	2·918878	− 0·613201	0·00006
6	2·050848	2·918889	− 0·613196	0·00000
7	2·050847	2·918886	− 0·613196	

Note that working accuracy was progressively increased.
Solution to 4D: $x = 2 \cdot 0508$, $y = 2 \cdot 9189$, $z = -0 \cdot 6132$.

ii) $x = 0 \cdot 1124$, $y = 0 \cdot 1236$, $z = 0 \cdot 1236$, $w = 0 \cdot 1124$.

STEP 14

N.B.   Although answers are given here, the student should not look
at them until he has carried out the suggested checks on his
own. The answers are given to 3D.

a)   (*full solution*)

m	$\mathbf{A}$			$\mathbf{I}$			Check	Row operation
	2	6	4	1	0	0	13	(1)
	6	19	12	0	1	0	38	(2)
	2	8	14	0	0	1	25	(3)
	2	6	4	1	0	0	13	(4) = (1)
-3	0	1	0	-3	1	0	-1	(5) = (2) − 3(1)
-1	0	2	10	-1	0	1	12	(6) = (3) − 1(1)
	2	6	4	1	0	0	13	(7) = (1)
	0	1	0	-3	1	0	-1	(8) = (5)
-2	0	0	10	5	-2	1	14	(9) = (6) − 2(5)
Inverse matrix			8·5	$-2 \cdot 6$	$-0 \cdot 2$			(Check that $\mathbf{A A}^{-1}$ = I)
			-3	1	0			
			0·5	$-0 \cdot 2$	0·1			

Note: The first column of $\mathbf{A}^{-1}$ is $\begin{pmatrix} u_3 \\ u_2 \\ u_1 \end{pmatrix}$, found by solving

$$\begin{bmatrix} 2 & 6 & 4 \\ 0 & 1 & 0 \\ 0 & 0 & 10 \end{bmatrix} \begin{bmatrix} u_3 \\ u_2 \\ u_1 \end{bmatrix} = \begin{bmatrix} 1 \\ -3 \\ 5 \end{bmatrix}$$ by back-substitution.

The second column is found by solving

$$\begin{bmatrix} 2 & 6 & 4 \\ 0 & 1 & 0 \\ 0 & 0 & 10 \end{bmatrix} \begin{bmatrix} u_6 \\ u_5 \\ u_4 \end{bmatrix} = \begin{bmatrix} 0 \\ 1 \\ -2 \end{bmatrix};$$

the third is from

$$\begin{bmatrix} 2 & 6 & 4 \\ 0 & 1 & 0 \\ 0 & 0 & 10 \end{bmatrix} \begin{bmatrix} u_9 \\ u_8 \\ u_7 \end{bmatrix} = \begin{bmatrix} 0 \\ 0 \\ 1 \end{bmatrix}.$$

b) $\begin{bmatrix} 0\cdot046 & -0\cdot605 & 1\cdot031 \\ 0\cdot448 & -0\cdot403 & 0\cdot398 \\ -0\cdot362 & +0\cdot851 & -1\cdot023 \end{bmatrix}$

c) $\begin{bmatrix} 0\cdot705 & 2\cdot544 & -2\cdot761 \\ -1\cdot371 & 0\cdot806 & 1\cdot609 \\ 2\cdot013 & -1\cdot808 & -0\cdot030 \end{bmatrix}$

2. a) $x = \begin{bmatrix} 25\cdot500 \\ -9\cdot000 \\ 2\cdot000 \end{bmatrix}, \begin{bmatrix} 2\cdot700 \\ -1\cdot000 \\ 0\cdot400 \end{bmatrix}$

b) $x = \begin{bmatrix} -4\cdot349 \\ -2\cdot448 \\ 5\cdot133 \end{bmatrix}, \begin{bmatrix} 0\cdot426 \\ -0\cdot005 \\ -0\cdot172 \end{bmatrix}$

c) $x = \begin{bmatrix} 6\cdot648 \\ 0\cdot103 \\ -1\cdot761 \end{bmatrix}, \begin{bmatrix} 2\cdot381 \\ 2\cdot937 \\ -2\cdot827 \end{bmatrix}$

STEP 15

1.

x	$f(x)=x^3$	First difference	Second	Third	Fourth
1	1				
		7			
2	8		12		
		19		6	
3	27		18		0
		37		6	
4	64		24		0
		61		6	
5	125		30		
		91			
6	216				

2.

x	$f(x) = e^x$	First difference	Second	Third	Fourth
0·1	1·105171				
		56663			
0·15	1·161834		2906		
		59569		147	
0·2	1·221403		3053		12
		62622		159	
0·25	1·284025		3212		4
		65834		163	
0·3	1·349859		3375		10
		69209		173	
0·35	1·419068		3548		9
		72757		182	
0·4	1·491825		3730		10
		76487		192	
0·45	1·568312		3922		
		80409			
0·5	1·648721				

There is just a hint of excessive 'noise' at the fourth differences.

STEP 16

1. i)   0·06263,   0·00320,   0·00018,   −0·00002.
    ii) 0·07275,   0·00354,   0·00016,   −0·00002.
    iii) 0·00338,   −0·00002.
    iv) 0·00306 in each case.
    v)   0·00016 in each case.

2. i)   Consider $f(x) = x$.

    ii) $\Delta^3 f_j = \Delta^2(\Delta f_j)$

    $= \Delta^2(f_{j+1} - f_j)$

    $= \Delta(\Delta f_{j+1} - \Delta f_j)$

    $= \Delta(f_{j+2} - 2f_{j+1} + f_j)$

    $= f_{j+3} - 3f_{j+2} + 3f_{j+1} - f_j.$

    iii) $\nabla^3 f_j = \nabla^2(\nabla f_j)$

    $= \nabla^2(f_j - f_{j-1})$

    $= \nabla(\nabla f_j - \nabla f_{j-1})$

    $= \nabla(f_j - 2f_{j-1} + f_{j-2})$

    $= f_j - 3f_{j-1} + 3f_{j-2} - f_{j-3}.$

iv) $\delta^3 f_j = \delta^2(\delta f_j)$

$= \delta^2(f_{j+\frac{1}{2}} - f_{j-\frac{1}{2}})$

$= \delta(\delta f_{j+\frac{1}{2}} - \delta f_{j-\frac{1}{2}})$

$= \delta(f_{j+1} - 2f_j + f_{j-1})$

$= f_{j+\frac{3}{2}} - 3f_{j+\frac{1}{2}} - 3f_{j-\frac{1}{2}} - f_{j-\frac{3}{2}}.$

STEP 17

1. i)

x	$f(x) = x^4$	Δ	Δ^2	Δ^3	Δ^4
0	0·0000				
		1			
0·1	0·0001		14		
		15		36	
0·2	0·0016		50		24
		65		60	
0·3	0·0081		110		24
		175		84	
0·4	0·0256		194		24
		369		108	
0·5	0·0625		302		24
		671		132	
0·6	0·1296		434		24
		1105		156	
0·7	0·2401		590		24
		1695		180	
0·8	0·4096		770		24
		2465		204	
0·9	0·6561		974		
		3439			
1·0	1·0000				

ii)

x	$f(x) = x^4$	Δ	Δ^2	Δ^3	Δ^4
0	0·000				
		0			
0·1	0·000		2		
		2		2	
0·2	0·002		4		6
		6		8	
0·3	0·008		12		−1
		18		7	
0·4	0·026		19		4
		37		11	
0·5	0·063		30		2
		67		13	
0·6	0·130		43		4
		110		17	
0·7	0·240		60		−1
		170		16	
0·8	0·410		76		6
		246		22	
0·9	0·656		98		
		344			
1·0	1·000				

Worst round-off error is $6\cdot0 - 2\cdot4 = 3\cdot6$.

2.

x	$f(x)$	Δ	Δ^2	Δ^3
0	3			
		−1		
1	2		6	
		5		6
2	7		12	
		17		6
3	24		18	
		35		6
4	59		24	
		59		
5	118			

Data fitted by a cubic.

STEP 18

1.

x	$f(x)$	Δ	Δ^2	Δ^3
2	3·0671			
		33417		
3	6·4088		752	
		34169		6
4	9·8257		758	
		34927		6
5	13·3184		764	
		35691		36
6	16·8875		800	
		36491		−84
7	20·5366		716	
		37207		96
8	24·2573		812	
		38019		−24
9	28·0592		788	
		38807		6
10	31·9399		794	
		39601		6
11	35·9000		800	
		40401		6
12	39·9401		806	
		41207		
13	44·0608			

Since the function is a cubic, we expect Δ^3 to be constant. We use any one of $36 = 6 + \varepsilon$, $-84 = 6 - 3\varepsilon$, $96 = 6 + 3\varepsilon$, $-24 = 6 - \varepsilon$ to deduce that $\varepsilon = 30$ and hence

$$f(7) = 20·5366 - 0·0030 = 20·5336$$

which indicates a mistake with repeated digits (see Step 2).

2.

x	$f(x)$	Δ	Δ^2	Δ^3
0	1·3246			
		785		
1	1·4031		−9	
		776		−27
2	1·4807		−36	
		740		81
3	1·5547		45	
		785		−80
4	1·6332		−35	
		750		26
5	1·7082		−9	
		741		22
6	1·7823		13	
		754		−62
7	1·8577		−49	
		705		62
8	1·9282		13	
		718		−20
9	2·0000		−7	
		711		−1
10	2·0711		−8	
		703		
11	2·1414			

The mistakes are obviously at $x = 3$ and $x = 7$. Using the same approach as in the second example above we estimate $\varepsilon_1 \approx \dfrac{-81-80}{6}$

≈ -27 and $\varepsilon_2 \approx \dfrac{62+62}{6} \approx 21$.

The suggested corrections are

$\qquad f(3) = 1·5547 + 27 = 1·5574$ (transposed digits see step 2)

and $\quad f(7) = 1·8577 - 21 = 1·8556$.

In the latter case $f(7) = 1·8557$ seems more likely (repeated digits).

The corrected third differences would then be $0, 0, 1, -1, 2, -2, 2, 0, -1$.

STEP 19

1. Difference table:

x	$f(x) = \cos x$	Δ	Δ^2
80°	0·1736		
		− 28	
80° 10′	0·1708		−1
		− 29	
80° 20′	0·1679		0
		− 29	
80° 30′	0·1650		1
		− 28	
80° 40′	0·1622		−1
		− 29	
80° 50′	0·1593		

i) $\cos 80° 35' = f(80° 30') + 0·5\ \Delta f(80° 30')$
$$= 0·1650 + 0·5\,(-0·0028) = 0·1636.$$

ii) $\cos 80° 35' = 0·1650\ +\ 0·5(-0·0028)$
$$+\ (0·5)\,(0·5)\,(-0·5)\,(-0·0001)$$
$$= 0·1636.$$

(The second order correction is $+0·0000125$.)

2. Difference table:

x	$f(x) = \tan x$	Δ	Δ^2	Δ^3
80°	5·671			
		98		
80° 10′	5·769		4	
		102		−1
80° 20′	5·871		3	
		105		0
80° 30′	5·976		3	
		108		2
80° 40′	6·084		5	
		113		
80° 50′	6·197			

The *second order* differences are approximately constant, so that *quadratic* approximation is appropriate: setting $\theta = \frac{1}{2}$,

$$\tan 80° 35' \approx f(80° 30') + 0·5\ \Delta f(80° 30') - (0·5)^3\ \Delta^2 f(80° 30').$$
$$= 5·976 + 0·5(0·108) - (0·5)^3(0·005) = 6·029.$$

STEP 20

1.

x	$f(x)=e^x$	Δ	Δ^2	Δ^3
0·10	1·10517			
		5666		
0·15	1·16183		291	
		5957		15
0·20	1·22140		306	
		6263		14
0·25	1·28403		320	
		6583		18
0·30	1·34986		338	
		6921		16
0·35	1·41907		354	
		7275		
0·40	1·49182			

i) $e^{0.14} = f(0·14) \approx f(0·1) + \frac{4}{5}(0·05666)$

$$+ \frac{1}{2}\frac{4}{5}\left(-\frac{1}{5}\right)(0·00291) + \frac{1}{6}\frac{4}{5}\left(-\frac{1}{5}\right)\left(-\frac{6}{5}\right)(0·00015)$$

$$= 1·10517 + 0·04532(8) - 0·00023(3) + 0·00000(5)$$

$$= 1·15027.$$

ii) $e^{0.315} = f(0·315) \approx f(0·30) + \frac{3}{10}(0·06583)$

$$+ \frac{1}{2}\frac{3}{10}\frac{13}{10}(0·00320) + \frac{1}{6}\frac{3}{10}\frac{13}{10}\frac{23}{10}(0·00014)$$

$$= 1·34986 + 0·01974(9) + 0·00062(4) + 0·00002(1)$$

$$= 1·37025.$$

2. The relation obviously holds for $j = 0$, and for $j = 1$ since
$$\Delta f(x_0) = f(x_0+h) - f(x_0) \Rightarrow f(x_0+h) = (1+\Delta)f(x_0).$$

We proceed to a 'proof by induction'; suppose the relation holds for $j = k$, so that
$$f_k = f_0 + k\Delta f_0 + \frac{k(k-1)}{2}\Delta^2 f_0 + \dots + \Delta^k f_0,$$
where $f_j = f(x_0+jh)$ as usual.

Then

$$\Delta f_k = \Delta f_0 + k\Delta^2 f_0 + \frac{k(k-1)}{2}\Delta^3 f_0 + \ldots + \Delta^{k+1} f_0,$$

but

$$\Delta f_k = f_{k+1} - f_k,$$

so that

$$
\begin{aligned}
f_{k+1} &= f_k + \Delta f_k \\
&= f_0 + (k+1)\Delta f_0 + \left\{\frac{k(k-1)}{2} + k\right\}\Delta^2 f_0 + \ldots + \Delta^{k+1} f_0 \\
&= f_0 + (k+1)\Delta f_0 + \frac{(k+1)k}{2}\Delta^2 f_0 + \ldots + \Delta^{k+1} f_0 ;
\end{aligned}
$$

i.e., the relation holds for $j = k+1$. We conclude that it holds for $j = 0, 1, 2, \ldots,$.

With reference to Section 4 of Step 20, note that

$$f_j = f(x_j) = P_n(x_j),$$

on setting

$$\theta = j = 0, 1, 2, \ldots, .$$

3. The relevant difference table is given in Answer 2 of Step 17. Since $f_0 = 3$, $\Delta f_0 = -1$, $\Delta^2 f_0 = 6$, $\Delta^3 f_0 = 6$, $\Delta^4 f_0 = 0$,

$$
\begin{aligned}
P(x) &= f_0 + \theta\Delta f_0 + \frac{\theta(\theta-1)}{2}\Delta^2 f_0 + \frac{\theta(\theta-1)(\theta-2)}{6}\Delta^3 f_0 \\
&= 3 - \theta + 3\theta(\theta-1) + \theta(\theta-1)(\theta-2) \\
&= \theta^3 - 2\theta + 3.
\end{aligned}
$$

Now $x = x_0 + \theta h = \theta$ (since $x_0 = 0$, $h = 1$) so that *the* collocation polynomial for the first four tabular entries is

$$P_3(x) = x^3 - 2x + 3.$$

The student may verify that any four adjacent tabular points have the same collocation cubic; this coincidence corresponds to

$$f(x) \equiv P_3(x);$$

i.e., the tabular function itself is a cubic, as indicated by the (exactly) constant third order differences.

STEP 21

The relevant difference table is given in the answer to Question 1 of Step 20.

i) Stirling's formula; $\theta = \frac{1}{5}$.

$$f(0\cdot31) = 1\cdot34986 + \tfrac{1}{5}\tfrac{1}{2}(0\cdot06583 + 0\cdot06921)$$
$$+ \tfrac{1}{2}\tfrac{1}{25}(0\cdot00338) + \tfrac{1}{6}\tfrac{6}{5}\tfrac{1}{5}(-\tfrac{4}{5})\tfrac{1}{2}(0\cdot00018 + 0\cdot00016)$$
$$= 1\cdot34986 + 0\cdot01350(4) + 0\cdot00006(8) - 0\cdot00000(5)$$
$$= 1\cdot36343.$$

ii) Everett's formula; $\theta = \dfrac{1}{5}, \ \bar{\theta} = \dfrac{4}{5}$.

$$f(0\cdot31) = \tfrac{4}{5}(1\cdot34986) + \tfrac{1}{6}\tfrac{9}{5}\tfrac{4}{5}(-\tfrac{1}{5})(0\cdot00338) + \tfrac{1}{5}(1\cdot41907)$$
$$+ \tfrac{1}{6}\tfrac{6}{5}\tfrac{1}{5}(-\tfrac{4}{5})(0\cdot00354)$$
$$= 1\cdot07988(8) - 0\cdot00016(2) + 0\cdot28381(4) - 0\cdot00011(3)$$
$$= 1\cdot36343.$$

iii) Bessel's formula: $\theta = 0\cdot3$.

$$f(0\cdot315) = \tfrac{1}{2}(1\cdot34986 + 1\cdot41907) + (-0\cdot2)(0\cdot06921)$$
$$+ \tfrac{1}{2}(0\cdot3)(-0\cdot7)\tfrac{1}{2}(0\cdot00338 + 0\cdot00354)$$
$$+ \tfrac{1}{6}(0\cdot3)(-0\cdot7)(-0\cdot2)(0\cdot00016)$$
$$= 1\cdot38446(5) - 0\cdot01384(2) - 0\cdot00036(3) + 0\cdot00000(1)$$
$$= 1\cdot37026.$$

iv) Everett's formula; $\theta = 0\cdot3, \ \bar{\theta} = 0\cdot7$.

$$f(0\cdot315) = (0\cdot7)(1\cdot34986) + \tfrac{1}{6}(1\cdot7)(0\cdot7)(-0\cdot3)(0\cdot00338)$$
$$+ (0\cdot3)(1\cdot41907) + \tfrac{1}{6}(1\cdot3)(0\cdot3)(-0\cdot7)(0\cdot00354)$$
$$= 0\cdot94490(2) - 0\cdot00020(1) + 0\cdot42572(1) - 0\cdot00016(1)$$
$$= 1\cdot37026.$$

STEP 22

The Lagrange coefficients are

$$L_0(x) = \frac{(x+1)(x-1)(x-3)(x-4)}{(-1)(-3)(-5)(-6)} \qquad \text{for } x_0 = -2,$$

$$L_1(x) = \frac{(x+2)(x-1)(x-3)(x-4)}{1(-2)(-4)(-5)} \qquad \text{for } x_1 = -1,$$

$$L_2(x) = \frac{(x+2)(x+1)(x-3)(x-4)}{3 \times 2(-2)(-3)} \qquad \text{for } x_2 = 1,$$

$$L_3(x) = \frac{(x+2)(x+1)(x-1)(x-4)}{5 \times 4 \times 2(-1)} \quad \text{for } x_3 = 3,$$

$$L_4(x) = \frac{(x+2)(x+1)(x-1)(x-3)}{6 \times 5 \times 3 \times 1} \quad \text{for } x_4 = 4.$$

Thus

$$f(0) = L_0(0) \times 46 + L_1(0) \times 4 + L_2(0) \times 4 + L_3(0) \times 156$$
$$+ L_4(0) \times 484$$
$$= (-92 + 36 + 40 - 468 + 484)/15$$
$$= 0.$$

STEP 23

1. Let us order the points such that $x_0 = 27$, $x_1 = 8$, $x_2 = 1$, $x_3 = 0$, $x_4 = 64$, to get the divided difference table (entries multiplied by 10^5):

x	$f(x)$				
27	3·00000				
		5263			
8	2·00000		−347		
		14286		384	
1	1·00000		−10714		−6
		100000		165	
0	0·00000		−1488		
		6250			
64	4·00000				

From Newton's formula:
$$f(20) = f(27) + (-7)f(27,8) + (-7)(+12)f(27,8,1)$$
$$+ (-7)(+12)(+19)f(27,8,1,0)$$
$$+ (-7)(+12)(+19)(+20)f(27,8,1,0,64)$$
$$= 3 - 7(0·05263) - 84(-0·00347)$$
$$- 1596(0·00384) - 31920(-0·00006)$$
$$= 3 - 0·36841 + 0·29148 - 6·12864 + 1·91520$$
$$= -1·29037(!)$$

Since the terms are *not* decreasing we cannot have much confidence in this result. The student may recall that this example was quoted in Section 3 of Step 22, in a warning concerning the use of the Lagrange interpolation formula in practice. With divided differences, we can at least see that interpolation for $f(20)$ is invalid!

2. i) Let us order the points such that $x_0 = -1, x_1 = 1, x_2 = -2,$ $x_3 = 3, x_4 = 4$, to get the divided difference scheme:

	x	$f(x)$					
$k = 0$	-1	4					
			0				
$k = 1$	1	4		14			
			-14		1		
$k = 2$	-2	46		18		2	
			22		11		
$k = 3$	3	156		51			
			328				
$k = 4$	4	484					

Then
$$\begin{aligned}
f(0) &= f(-1) + (+1)f(-1,1) + (+1)(-1)f(-1,1,-2) \\
&\quad + (+1)(-1)(+2)f(-1,1,-2,3) \\
&\quad + (+1)(-1)(+2)(-3)f(-1,1,-2,3,4) \\
&= 4 + 1 \times 0 - 1 \times 14 - 2 \times 1 + 6 \times 2 \\
&= 0.
\end{aligned}$$

ii) Let us again order the points such that $x_0 = -1$, $x_1 = 1$, $x_2 = -2$, $x_3 = 3$, $x_4 = 4$, to get the Aitken scheme:

	x	$f(x)$					$x_k - x$
$k = 0$	-1	4					-1
$k = 1$	1	4	4				1
$k = 2$	-2	46	-38	-10			-2
$k = 3$	3	156	42	-15	-12		3
$k = 4$	4	484	100	-28	-16	0	4

The validity of this interpolation is dubious. The terms in Newton's divided difference formula (c.f. (i)) are *not* decreasing notably; in the Aitken scheme, we do *not* obtain a repeated value on the diagonal.

3.

x	$f(x)$				$x_k - x$
1	2·3919				−1
3	2·3938	2·3928(5)			+1
0	2·3913	2·3925	2·3927(3)		−2
4	2·3951	2·3929(7)	2·3927(3)	2·3927(3)	+2

STEP 24

1. The root of $f(x) = x + \cos x$ is in the interval $-0.8 < x < -0.7$; in fact,

$$f(-0.8) = -0.1033$$

and

$$f(-0.7) = +0.0648.$$

Since $f(x)$ is known explicitly, one may readily subtabulate (by successive interval bisection, say) and use linear inverse interpolation:

$$f(-0.75) = -0.0183, \quad \theta = \frac{0 + 0.0183}{0.0648 + 0.0183} = 0.2202,$$

whence $x = -0.75 + (0.2202)(0.05) = -0.7390$;

$$f(-0.725) = +0.0235, \quad \theta = \frac{0 + 0.0183}{0.0235 + 0.0183} = 0.4378,$$

whence $x = -0.75 + (0.4378)(0.025) = -0.7391$;

$$f(-0.7375) = +0.0027, \quad \theta = \frac{0 + 0.0183}{0.0027 + 0.0183} = 0.8714,$$

whence $x = -0.75 + (0.8714)(0.0125) = -0.7391$.

Checking $x = -0.7391$, we have $f(-0.7391) = 0.0000$.

2. The function $f(x) = 3xe^x$ increases as x increases, so that there is a unique x for $f(x) = 1$. Indeed, in Step 7 we noted that $0.25 < x < 0.27$, and this interval is quite small enough for linear inverse interpolation: since $f(0.27) = 1.0611$ and $f(0.25) = 0.9630$, we have

$$\theta = \frac{1.0000 - 0.9630}{1.0611 - 0.9630} = 0.3772$$

whence $x = 0.25 + (0.3772)(0.02) = 0.2575$.

Checking $x = 0.258$, we have $f(0.258) = 1.0018$, which is closer to $f(x) = 1$ than $f(0.257) = 0.9969$. (While the value to $3D$ is obtained immediately by linear inverse interpolation, the method of bisection described in Step 7 may be preferred when greater accuracy is demanded.)

3. If the explicit form of the function is unknown so that it is not possible to subtabulate readily, one may use iterative inverse interpolation. The relevant difference table is:

x	$f(x)$	Δf	$\Delta^2 f$	$\Delta^3 f$
2	3·0671			
		33417		
3	6·4088		752	
		34169		6
4	9·8257		758	
		34927		6
5	13·3184		764	
		35691		6
6	16·8875		770	
		36461		6
7	20·5336		776	
		37237		6
8	24·2573		782	
		38019		6
9	28·0592		788	
		38807		6
10	31·9399		794	
		39601		6
11	35·9000		800	
		40401		6
12	39·9401		806	
		41207		
13	44·0608			

To find x for $f(x) = 10$, one may use inverse interpolation based on Newton's forward formula:

$$\theta_1 = (10 - 9.8257)/3.4927 = 0.1743/3.4927 = 0.04990 \approx 0.05$$

$$\theta_2 = \{0.1743 - \tfrac{1}{2}(0.05)(-0.95)(0.0764)\}/3.4927$$

$$= \{0.1743 + 0.0018\}/3.4927 = 0.0504$$

and further corrections are negligible, so that

$$x = 4 + 0.0504 = 4.0504.$$

To find x for $f(x) = 20$, one may choose inverse interpolation based on Everett's formula:

$$\theta_1 = (20 - 16{\cdot}8875)/3{\cdot}6461 = 3{\cdot}1125/3{\cdot}6461 = 0{\cdot}85365 \approx 0{\cdot}85$$

$$\theta_2 = \{3{\cdot}1125 + \tfrac{1}{6}(0{\cdot}85)\,(-\,0{\cdot}15)[(-\,1{\cdot}15)\,(0{\cdot}0770)$$
$$-\,(1{\cdot}85)\,(0{\cdot}0776)]\}/3{\cdot}6461$$
$$= \{3{\cdot}1125 + (0{\cdot}02125)\,(0{\cdot}2321)\}/3{\cdot}6461$$
$$= \{3{\cdot}1125 + 0{\cdot}0049\}/3{\cdot}6461 = 0{\cdot}8550$$

and further corrections are negligible, so that
$$x = 6 + 0{\cdot}8550 = 6{\cdot}8550.$$
(Of course, the student may prefer to use the form based on Newton's forward formula.)

To find x for $f(x) = 40$, one may choose inverse interpolation based on Newton's *backward* formula. Thus,

$$\theta_1 = \{f(x) - f_j\}/\nabla f_j,$$
$$\theta_2 = \{f(x) - f_j - \tfrac{1}{2}\theta_1(\theta_1 + 1)\nabla^2 f_j\}/\nabla f_j,$$

etc. Consequently,

$$\theta_1 = (40 - 39{\cdot}9401)/4{\cdot}0401 = 0{\cdot}0599/4{\cdot}0401 = 0{\cdot}0148 \approx 0{\cdot}015$$

$$\theta_2 = \{0{\cdot}0599 - \tfrac{1}{2}(0{\cdot}015)\,(1{\cdot}015)\,(0{\cdot}0800)\}/4{\cdot}0401$$

$$= \{0{\cdot}0599 - 0{\cdot}0006\}/4{\cdot}0401 = 0{\cdot}0147$$

and further corrections are negligible, so that

$$x = 12 + 0{\cdot}0147 = 12{\cdot}0147.$$

Let us now consider the check by direct interpolation. We have from Newton's forward formula

$$f(4{\cdot}0504) = 9{\cdot}8257 + (0{\cdot}0504)\,(3{\cdot}4927)$$
$$+ \tfrac{1}{2}(0{\cdot}0504)\,(-0{\cdot}9496)\,(0{\cdot}0764)$$
$$= 10{\cdot}00$$

and

$$f(6{\cdot}8550) = 16{\cdot}8875 + (0{\cdot}8550)\,(3{\cdot}6461)$$
$$+ \tfrac{1}{2}(0{\cdot}8550)\,(-0{\cdot}1450)\,(0{\cdot}0776)$$
$$= 20{\cdot}00,$$

while from Newton's backward formula

$$f(12 \cdot 0147) = 39 \cdot 9401 + (0 \cdot 0147)(4 \cdot 0401)$$
$$+ \tfrac{1}{2}(0 \cdot 0147)(1 \cdot 0147)(0 \cdot 0800).$$

$$= 40 \cdot 00.$$

Finally, we may determine the cubic $f(x)$ and use it to check the answers:

$$f(x) = f_j + \theta \Delta f_j + \tfrac{1}{2}\theta(\theta-1)\Delta^2 f_j + \tfrac{1}{6}\theta(\theta-1)(\theta-2)\Delta^3 f_j$$
$$= 9 \cdot 8257 + (x-4)(3 \cdot 4927) + \tfrac{1}{2}(x-4)(x-5)(0 \cdot 0764)$$
$$+ \tfrac{1}{6}(x-4)(x-5)(x-6)(0 \cdot 0006)$$
$$= \{9 \cdot 8257 - 4(3 \cdot 4927) + 10(0 \cdot 0764) - 20(0 \cdot 0006)\}$$
$$+ \{3 \cdot 4927 - \tfrac{9}{2}(0 \cdot 0764) + \tfrac{37}{3}(0 \cdot 0006)\}x$$
$$+ \{\tfrac{1}{2}(0 \cdot 0764) - \tfrac{5}{2}(0 \cdot 0006)\}x^2 + \tfrac{1}{6}(0 \cdot 0006)x^3$$
$$= 0 \cdot 0001 x^3 + 0 \cdot 0367 x^2 + 3 \cdot 1563 x - 3 \cdot 3931,$$

hence

$$f(4 \cdot 0504) = 9 \cdot 9999,$$
$$f(6 \cdot 8550) = 20 \cdot 0001,$$
$$f(12 \cdot 0147) = 40 \cdot 0001.$$

(In each case, the value obtained by iterative inverse interpolation in fact renders the corresponding function value accurate to 3D.)

STEP 25

1. The following table displays the line and parabola values for y, the respective errors, and squared errors.

Line equation: $y = 2 \cdot 13 + 0 \cdot 2x$
Parabola equation: $y = -1 \cdot 2 + 2 \cdot 7x - 0 \cdot 36x^2$

x	1	2	3	4	5	6		
y	1	3	4	3	4	2		
l = line y-value	2·33	2·53	2·73	2·93	3·13	3·33		
line error $(y - l)$	−1·33	0·47	1·27	0·07	0·87	−1·33		
$(y - l)^2$	1·769	0·221	1·613	0·005	0·757	1·769	6·134	= S (line)
p = parabola y-value	1·14	2·76	3·66	3·84	3·30	2·04		
parabola error $(y - p)$	−0·14	0·24	0·34	−0·84	0·70	−0·04		
$(y - p)^2$	0·0196	0·0576	0·1156	0·7056	0·4900	0·0016	1·390	= S (parabola)

2. Computing n, Σx, Σy, Σx^2 and Σxy, inserting in the normal equations and solving gives:

   i) *Normal equations:*
   $$23·9 = 8c_1 + 348c_2,$$
   $$1049·1 = 348c_1 + 15260c_2.$$

   Least squares line: $y = -0·38 + 0·077x.$

   Prediction: % nickel (y) when $x = 38$ is 2·55 (to 2D).

   ii) *Normal equations:*
   $$348 = 6c_1 + 219c_2,$$
   $$13659 = 219c_1 + 8531c_2.$$

   Least squares line: $y = -6·99 + 1·78x.$

   Prediction: sales (y) when $x = 48$ is 78·45 ($\times$ \$100).

4. The matrix form for the normal equations (see question 3) is

   $$\begin{bmatrix} 9 \\ 24 \\ 72 \end{bmatrix} = \begin{bmatrix} 5 & 10 & 30 \\ 10 & 30 & 100 \\ 30 & 100 & 354 \end{bmatrix} \begin{bmatrix} c_1 \\ c_2 \\ c_3 \end{bmatrix},$$
   the elements being the sums $\Sigma y = 9$, $\Sigma xy = 24$, etc.

   Solution: $y = c_1 + c_2 x + c_3 x^2 = -0·2572 + 2·3144x - 0·4286x^2$
   $S = 0·6286$ to 4D.

5. We require to find c_1 and c_2 to minimize
   $$S = \sum_{i=1}^{4} \varepsilon^2_{\,i} = \sum_{i=1}^{4} (y_i - c_1 - c_2 \sin x_i)^2.$$
   Now $\dfrac{\partial S}{\partial c_1} = \sum -2(y_i - c_1 - c_2 \sin x_i)$

and $\dfrac{\partial S}{\partial c_2} = \sum -2 \, (y_i - c_1 - c_2 \sin x_i) \sin x_i$

so normal equations may be written as

$$\sum y_i = 4c_1 + (\sum \sin x_i) \, c_2$$

and $\sum y_i \sin x_i = (\sum \sin x_i) \, c_1 + (\sum \sin^2 x_i) \, c_2.$

Tabulating:

x_i	y_i	$\sin x_i$	$y_i \sin x_i$	$\sin^2 x_i$
0	0	0	0	0
$\pi/6$	1	0·5	0·5	0·25
$\pi/2$	3	1	3	1
$5\pi/6$	2	0·5	1	0·25
$\sum$	6	2	4·5	1·5

Solving the equations

$$6 = 4c_1 + 2c_2$$
$$4{\cdot}5 = 2c_1 + 1{\cdot}5c_2$$

we obtain $c_1 = 0, c_2 = 3.$

STEP 26

1. Consider the Newton backward difference formula

$$f(x) = f(x_j + \theta h) \approx \left[1 + \theta\nabla + \tfrac{1}{2}\theta(\theta+1)\nabla^2 + \dfrac{\theta(\theta+1)(\theta+2)}{3!}\nabla^3 + \ldots\right]f_j$$

$$f'(x) = \frac{1}{h}\frac{df}{d\theta} \approx \frac{1}{h}\left[\nabla + (\theta+\tfrac{1}{2})\nabla^2 + \dfrac{3\theta^2 + 6\theta + 2}{6}\nabla^3 + \ldots\right]f_j$$

$$f''(x) = \frac{1}{h^2}\frac{d^2f}{d\theta^2} \approx \frac{1}{h^2}\left[\nabla^2 + (\theta+1)\nabla^3 + \ldots\right]f_j$$

2.   Difference table:

x	$f(x)$	Δ	Δ^2	Δ^3
1·00	1·00000			
		2470		
1·05	1·02470		− 59	
		2411		5
1·10	1·04881		− 54	
		2357		4
1·15	1·07238		− 50	
		2307		1
1·20	1·09545		− 49	
		2258		6
1·25	1·11803		− 43	
		2215		
1·30	1·14018			

Note that $h = 0·05$: thus,

     i)

$$f'(1·00) \approx 20[\Delta - \tfrac{1}{2}\Delta^2 + \tfrac{1}{3}\Delta^3]f(1·00)$$

$$= 20\,(0·02470 + 0·000295 + 0·000017)$$

$$= 0·50024.$$

$$f''(1·00) \approx (20)^2[\Delta^2 - \Delta^3]f(1·00)$$

$$= 400(-0·00059 - 0·00005)$$

$$= -0·256.$$

The correct values are of course

$$f'(1·00) = \left.\frac{1}{2\sqrt{x}}\right|_{x\,=\,1·00} = 0·5$$

$$f''(1·00) = \left.-\frac{1}{4x^{3/2}}\right|_{x\,=\,1·00} = -0·25\,;$$

although the input data are correct to $5D$, the results are accurate only to $3D$ and $1D$ respectively.

     ii)

$$f'(1·30) \approx 20\left[\nabla + \frac{1}{2}\nabla^2 + \frac{1}{3}\nabla^3\right]f(1·30)$$

$$= 20(0·02215 - 0·00021(5) + 0·00002)$$

$$= 0·4391.$$

$$f''(1{\cdot}30) \approx (20)^2 [\nabla^2 + \nabla^3] f(1{\cdot}30)$$
$$= 400(-0{\cdot}00043 + 0{\cdot}00006)$$
$$= -0{\cdot}148.$$

To 4D the correct values are $0{\cdot}4385$ and $-0{\cdot}1687$.

3. i) Expanding about $x = x_j$:

$$f(x_j + h) = f(x_j) + hf'(x_j) + \tfrac{1}{2}h^2 f''(x_j) + \ldots,$$

so $(f(x_j + h) - f(x_j))/h = f'(x_j) + \tfrac{1}{2}hf''(x_j) + \ldots$

and the error $\approx \tfrac{1}{2}hf''(x_j)$.

ii) Expanding about $x = x_j + \tfrac{1}{2}h$:

$$f(x_j + h) = f(x_j + \tfrac{1}{2}h) + \tfrac{1}{2}hf'(x_j + \tfrac{1}{2})$$
$$+ \tfrac{1}{8}h^2 f''(x_j + \tfrac{1}{2}h) + \tfrac{1}{48}h^3 f'''(x_j + \tfrac{1}{2}h) + \ldots$$

and $f(x_j) = f(x_j + \tfrac{1}{2}h) - \tfrac{1}{2}hf'(x_j + \tfrac{1}{2}h) + \tfrac{1}{8}h^2 f''(x_j + \tfrac{1}{2}h)$
$$- \tfrac{1}{48}h^3 f'''(x_j + \tfrac{1}{2}h) + \ldots$$

so $(f(x_j + h) - f(x_j))/h = f'(x_j + \tfrac{1}{2}h) + \tfrac{1}{24}h^2 f'''(x_j + \tfrac{1}{2}h) + \ldots$

and the error $\approx \tfrac{1}{24}h^2 f'''(x_j + \tfrac{1}{2}h)$.

iii) Expanding about $x = x_j$:

$$f(x_j + 2h) = f(x_j) + 2hf'(x_j) + 2h^2 f''(x_j) + \tfrac{4}{3}h^3 f'''(x_j) + \ldots,$$

so $(f(x_j + 2h) - 2f(x_j + h) + f(x_j))/h^2 = f''(x_j) + hf'''(x_j) + \ldots$

and the error $\approx hf'''(x_j)$.

iv) Expanding about $x = x_j + h$:

$$f(x_j + 2h) = f(x_j + h) + hf'(x_j + h) + \tfrac{1}{2}h^2 f''(x_j + h)$$
$$+ \tfrac{1}{6}h^3 f'''(x_j + h) + \tfrac{1}{24}h^4 f^{iv}(x_j + h) + \ldots$$

$f(x_j) = f(x_j + h) - hf'(x_j + h) + \tfrac{1}{2}h^2 f''(x_j + h) - \tfrac{1}{6}h^3 f'''(x_j + h)$
$$+ \tfrac{1}{24}h^4 f^{iv}(x_j + h) + \ldots$$

so $(f(x_j + 2h) - 2f(x_j + h) + f(x_j))/h^2 = f''(x_j + h)$
$$+ \tfrac{1}{12}h^2 f^{iv}(x_j + h) + \ldots$$

and the error $\approx \tfrac{1}{12}h^2 f^{iv}(x_j + h)$.

STEP 27

1. With $b - a = 1 \cdot 30 - 1 \cdot 00 = 0 \cdot 30$, we may choose
    $$h = 0 \cdot 30, 0 \cdot 15, 0 \cdot 10, 0 \cdot 05, \dots .$$

If $T(h)$ denotes the approximation corresponding to strip width h, we get

$$T(0 \cdot 30) = \frac{0 \cdot 30}{2} (1 \cdot 00000 + 1 \cdot 14018) = 0 \cdot 32102(7),$$

$$T(0 \cdot 15) = \frac{0 \cdot 15}{2} (1 \cdot 00000 + 1 \cdot 14018) + (0 \cdot 15)(1 \cdot 07238)$$

$$= 0 \cdot 16051(4) + 0 \cdot 16085(7) = 0 \cdot 32137(1),$$

$$T(0 \cdot 10) = \frac{0 \cdot 10}{2} (1 \cdot 00000 + 1 \cdot 14018) + (0 \cdot 10)(1 \cdot 04881 + 1 \cdot 09545)$$

$$= 0 \cdot 10700(9) + 0 \cdot 21442(6) = 0 \cdot 32143(5),$$

$$T(0 \cdot 05) = \frac{0 \cdot 05}{2} (1 \cdot 0000 + 1 \cdot 14018) +$$

$$+ (0 \cdot 05)(1 \cdot 02470 + 1 \cdot 04881 + 1 \cdot 07238 + 1 \cdot 09545$$
$$+ 1 \cdot 11803)$$

$$= 0 \cdot 05350(5) + 0 \cdot 26796(9) = 0 \cdot 32147(4)$$

To $8D$, the answer is in fact $0 \cdot 32148537$, so that we may observe that the error sequence $0 \cdot 00045(8)$, $0 \cdot 00011(4)$, $0 \cdot 00005(0)$, $0 \cdot 00001(1)$ corresponds to a decrease with h^2 (the truncation error dominates round-off).

2. $T(1) = \frac{1}{2} \{ \frac{1}{1 + 0} + \frac{1}{1 + 1} \} = 0 \cdot 75,$

$$T(0 \cdot 5) = \frac{0 \cdot 5}{2} \{ \frac{1}{1 + 0} + \frac{1}{1 + 1} \} + 0 \cdot 5 (\frac{1}{1 + 0 \cdot 5}) = 0 \cdot 7083 \text{ (to } 4D),$$

$$T(0 \cdot 25) = \frac{0 \cdot 25}{2} \{ \frac{1}{1 + 0} + \frac{1}{1 + 1} \} + 0 \cdot 25 \{ \frac{1}{1 + 0 \cdot 25} + \frac{1}{1 + 0 \cdot 5}$$

$$+ \frac{1}{1 + 0 \cdot 75} \} = 0 \cdot 6970 \text{ (to } 4D).$$

The correct value is $\log_e 2 \approx 0 \cdot 6931$, so the errors are (approximately) $0 \cdot 0569$, $0 \cdot 0152$ and $0 \cdot 0039$ respectively (note the decrease with h^2).

STEP 28

We have

$$f(x) = \frac{1}{1+x},$$

$$f''(x) = \frac{2}{(1+x)^3},$$

$$f^{(4)}(x) = \frac{24}{(1+x)^5}.$$

The truncation error bound for the trapezoidal rule is $\frac{2}{12}h^2 = \frac{1}{6}h^2$, so that we would need to choose $h \leqslant 0.01$ to obtain $4D$ accuracy. For Simpson's rule, however, the truncation error bound is $\frac{24}{180}h^4 = \frac{2}{15}h^4$: we may choose $h = 0.1$. Tabulating:

x	0	0·1	0·2	0·3	0·4	0·5
$f(x)$	1·000000	0·909091	0·833333	0·769231	0·714286	0·666667

x	0·6	0·7	0·8	0·9	1·0
$f(x)$	0·625000	0·588235	0·555556	0·526316	0·500000

By Simpson's rule,

$$\int_0^1 \frac{1}{1+x}\,dx = \frac{0.1}{3}[1 + 4(0.909091 + 0.769231 + 0.666667$$
$$+ 0.588235 + 0.526316)$$
$$+ 2(0.833333 + 0.714286 + 0.625000$$
$$+ 0.555556) + 0.500000]$$
$$= 0.6931(5).$$

(Note that working to at least $5D$ is necessary to guard against round-off error, and that $\int_0^1 \frac{1}{1+x}\,dx = 0.693147$ to $6D$.)

STEP 29

Refer to the difference table (see below):

i) $x_0 = 0.88$, $x_0 + h = 0.92$ (forward differences)

$$\int_{0.88}^{0.92} f(x)\,dx \approx 0.04\,(1.2097 + 0.0518 - 0.0009(6) + 0.0001)$$

$$= 0.04 (1.2606)$$
$$= 0.0504.$$

ii) $x_0 = 0.88$, $x_0 + h = 0.92$ (backward differences)

$$\int_{0.88}^{0.92} f(x)dx \approx 0.04 (1.2097 + 0.0470(5) + 0.0033(8)$$
$$+ 0.0004(5))$$
$$= 0.04 (1.2606)$$
$$= 0.0504.$$

The difference table is:

x	$f(x)$	Δ	Δ^2	Δ^3	Δ^4
0·68	0·8087				
		684			
0·72	0·8771		50		
		734		7	
0·76	0·9505		57		5
		791		12	
0·80	1·0296		69		0
		860		12	
0·84	1·1156		81		2
		941		14	
0·88	1·2097		95		6
		1036		20	
0·92	1·3133		115		4
		1151		24	
0·96	1·4284		139		9
		1290		33	
1·00	1·5574		172		9
		1462		42	
1·04	1·7036		214		16
		1676		58	
1·08	1·8712		272		
		1948			
1·12	2·0660				

STEP 30

Change of variable:
$$u = \frac{1}{2}(x + 1).$$

$$\int_0^1 \frac{1}{1+u} \, du = \frac{1}{2} \int_{-1}^1 \frac{1}{1 + \frac{1}{2}(x+1)} \, dx$$

$$= \int_{-1}^{1} \frac{dx}{3+x} \, .$$

Two-point formula:

$$\int_{0}^{1} \frac{1}{1+u} \, du \approx \frac{1}{3 - 0.57735027} + \frac{1}{3 + 0.57735027}$$

$$= 0.412771 + 0.279537$$

$$= 0.692308,$$

which is correct to $2D$.

Four point formula:

$$\int_{0}^{1} \frac{1}{1+u} \, du \approx 0.34785485 \left\{ \frac{1}{3 - 0.86113631} + \frac{1}{3 + 0.86113631} \right\}$$

$$+ 0.65214515 \left\{ \frac{1}{3 - 0.33998104} + \frac{1}{3 + 0.33998104} \right\}$$

$$\approx 0.347855 \, [0.467538 + 0.258991]$$
$$+ 0.652145 \, [0.375937 + 0.299403]$$
$$= (0.347855)(0.726529) + (0.652145)(0.675340)$$
$$= 0.252727 + 0.440420$$
$$= 0.693147 \text{ (correct to } 6D).$$

STEP 31

1. $y_{n+1} = y_{n-1} + 0.2(x_n + y_n),$

$y_2 \quad = 1 + 0.2(0.1 + 1.11) = 1.242,$

$y_3 \quad = 1.11 + 0.2(0.2 + 1.242) = 1.3984,$

$y_4 \quad = 1.242 + 0.2(0.3 + 1.3984) = 1.58168,$

and $y_5 \quad = 1.3984 + 0.2(0.4 + 1.58168) = 1.794736,$

which has an error of approximately 0.003 (compare with Section 4(c) of Step 31).

2. $y_{n+1} = y_n - 0.2x_n y_n^2 = y_n(1 - 0.2x_n y_n),$

$y_1 \quad = 2(1 - 0.2 \times 0 \times 2) = 2,$

$y_2 \quad = 2(1 - 0.2 \times 0.2 \times 2) = 1.84,$

$y_3 \quad = 1 \cdot 84(1 - 0 \cdot 2 \times 0 \cdot 4 \times 1 \cdot 84) = 1 \cdot 56915,$

$y_4 \quad = 1 \cdot 56915(1 - 0 \cdot 2 \times 0 \cdot 6 \times 1 \cdot 56915) = 1 \cdot 27368,$

$y_5 \quad = 1 \cdot 27368(1 - 0 \cdot 2 \times 0 \cdot 8 \times 1 \cdot 27368) = 1 \cdot 01412.$

The exact solution is $y(x) = \dfrac{2}{1+x^2}$, so $y(1) = 1$ and the error in y_5 is approximately 0·014.

INDEX

INDEX